REVISE OCR AS/A LEVEL
Physics
REVISION GUIDE

D0189131

Series Consultant: Harry Smith

Authors: Steve Adams and Ken Clays

Our revision resources are the smart choice for those revising for OCR AS/A Level Physics. This book will help you to:

- **Organise** your revision with the one-topic-per-page format
- **Speed up** your revision with summary notes in short, memorable chunks
- **Track** your revision progress with at-a-glance check boxes
- **Check** your understanding with worked examples
- **Develop** your exam technique with exam-style practice questions and full answers.

Revision is more than just this Guide!

Make sure that you have practised every topic covered in this book, with the accompanying OCR AS/A Level Physics A Revision Workbook. It gives you:

- More exam-style practice and a 1-to-1 page match with this Revision Guide
- Guided questions to help build your confidence
- Hints to support your revision and practice.

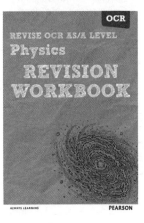

For the full range of Pearson revision titles across GCSE, A Level + BTEC:

www.pearsonschools.co.uk/revise

ALWAYS LEARNING

PEARSON

Contents

Quantities and units

You should always give units along with numerical values for physical quantities (except ratios).

S.I. base units

Quantity	Symbol	S.I. unit
mass	m	kg
length	l	m
time	t	s
current	I	A
temperature	T	K
amount of substance	none	mol

Note that kg and not g is the S.I. unit for mass!

Derived units

Many physical quantities are derived from base quantities and so their S.I. Système Internationale units are combinations of base units. Velocity is a good example: it is displacement ÷ time, so its unit is $m\,s^{-1}$. Some derived units have special names, but all of them can be reduced to combinations of base units.

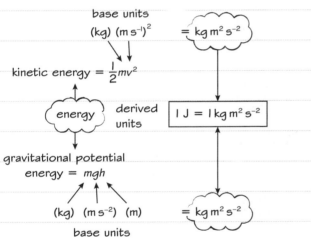

base units
(kg) $(m\,s^{-1})^2$ = $kg\,m^2\,s^{-2}$

kinetic energy = $\frac{1}{2}mv^2$

energy — derived units

$1\,J = 1\,kg\,m^2\,s^{-2}$

gravitational potential energy = mgh

(kg) $(m\,s^{-2})$ (m) = $kg\,m^2\,s^{-2}$

base units

All equations in physics must be balanced, or **homogeneous**, so units must be the same on the left- and right-hand sides of the equation.

Quantity	Symbol	Derived units	Special name
energy	E	$kg\,m^2\,s^{-2}$	J (joule)
force	F	$kg\,m\,s^{-2}$	N (newton)
power	P	$kg\,m^2\,s^{-3}$	W (watt)
density	ρ	$kg\,m^{-3}$	
pressure	p	$kg\,m^{-1}\,s^{-2}$	Pa (pascal)
velocity	u, v	$m\,s^{-1}$	
acceleration	a	$m\,s^{-2}$	
charge	Q	$A\,s$	C (coulomb)
potential difference	V	$kg\,m^2\,s^{-3}\,A^{-1}$	V (volt)
resistance	R	$kg\,m^2\,s^{-3}\,A^{-2}$	Ω (ohm)

Check that derived equations are homogeneous. If it is not then you have made a mistake!

Maths skills — Prefixes

Prefix		Multiplier
feonto	f	$\times 10^{-15}$
pico-	p	$\times 10^{-12}$
nano-	n	$\times 10^{-9}$
micro-	μ	$\times 10^{-6}$
milli-	m	$\times 10^{-3}$
centi-	c	$\times 10^{-2}$
deci-	d	$\times 10^{-1}$
kilo-	k	$\times 10^{3}$
mega-	M	$\times 10^{6}$
giga-	G	$\times 10^{9}$
tera-	T	$\times 10^{12}$

Prefixes make it easy to express large or small values. For example, 25 000 m is the same as 25 km. The prefix kilo- multiplies the 25 by 10^3. S.I. uses prefixes in steps of 10^3.

You will also see centi- and deci- used, though they are not part of the S.I.

Worked example

This equation is used to calculate the energy needed to raise the temperature of a certain mass of material.

$$E = mc\Delta T$$

The quantity c is called specific heat capacity. What are the derived units for specific heat capacity expressed in terms of base S.I. units? **(3 marks)**

Rearrange the equation to give $c = \dfrac{E}{m\Delta T}$.

The equation must be homogeneous, so c must have the same units as $\dfrac{E}{m\Delta T}$: $J\,kg^{-1}\,K^{-1}$.

Since $1\,J = 1\,kg\,m^2\,s^{-2}$

the unit for c is $kg\,m^2\,s^{-2} \times kg^{-1}\,K^{-1} = m^2\,s^{-2}\,K^{-1}$

Now try this

1. Newton discovered an equation for the gravitational force between two masses separated by a distance r.

$$F = \frac{Gm_1m_2}{r^2}$$

where G is the universal gravitational constant. Find a derived unit for G and express it in terms of base S.I. units. **(3 marks)**

2. A metal wire with circular cross section has a length of 10.0 cm and a radius of 0.45 mm. Calculate its volume in m^3. **(2 marks)**

Estimating physical quantities

Maths skills You should practice estimating the value of physical quantities to get a feel for sizes and likely answers – it is surprising how good these estimates can be, even when based on limited data.

Standard form...

...is a number between 1 and 10 multiplied by a power of 10. The power of 10 tells you how many places to move the decimal point.

speed of light $c = 3.0 \times 10^8 \, \text{m s}^{-1}$

$\underset{1\,2\,3\,4\,5\,6\,7\,8}{\underbrace{}}$

3.00000000
move decimal point 8 places
$c = 300\,000\,000 \, \text{m s}^{-1}$

charge on an electron $e = 1.6 \times 10^{-19} \, \text{C}$

1918171615141312111098765432 1
00000000000000000001.6
$e = 0.000\,000\,000\,000\,000\,000\,000\,16 \, \text{C}$

Equivalent ways of representing physical quantities

Proton diameter
- In S.I. base units:
  $0.000\,000\,000\,000\,000\,000\,88 \, \text{m}$
- Using a prefix: 0.88 fm
- Standard form: $8.8 \times 10^{-16} \, \text{m}$

Age of the Universe
- Non-S.I. units: 13.7 billion years
- In S.I. base units: $430\,000\,000\,000\,000\,000 \, \text{s}$
- Using a prefix: 430 000 Ts
- Standard form: $4.3 \times 10^{17} \, \text{s}$

Look at page 1 for a list of prefixes.

Order of magnitude

To make a rough comparison of the **magnitudes** (sizes) of physical quantities, use order of magnitude, usually expressed as a power of 10.

Quantity	Order of magnitude
nuclear radius	10^{-15} m
atomic radius	10^{-10} m
human height	10^1 m
Earth' radius	10^7 m
Earth' orbit	10^{11} m

Making estimates

 Identify the relevant physical quantities.

 Estimate the value of each quantity using your knowledge (not a guess!).

 Combine the quantities.

How many atoms in a house brick?

 You need to know the volume of a brick and the volume of an atom.

 Estimate brick volume:
30 cm × 15 cm × 10 cm = 4500 cm³
This is an estimate only so round to
5000 cm³ $= 5 \times 10^{-3} \, \text{m}^3$.

 Estimate atom volume:
$(10^{-10} \, \text{m})^3 = 10^{-30} \, \text{m}^3$

Number of atoms $N = \dfrac{\text{volume of brick}}{\text{volume of atom}}$

$N = \dfrac{5 \times 10^{-3}}{10^{-30}} = 5 \times 10^{27}$ atoms

Worked example

1 Express the mass of the Sun,
   1 989 000 000 000 000 000 000 000 000 000 kg, in
   standard form.　**(1 mark)**

The decimal point has to move 30 places to the left to give a number between 1 and 10. The mass of the Sun is therefore 1.989×10^{30} kg.

2 Estimate the time it would take to drive a car non-stop from the north to the south of South America
   (4 marks)

The speed of a car could be about 80 km h⁻¹ (50 miles/h). This is about 22 m s⁻¹ so use 20 m s⁻¹. The Earth's radius is about 10^7 m. The circumference of the Earth is about $2\pi \times 10^7$ m, and South America stretches over roughly half of one hemisphere, so the distance is about $(2\pi \times 10^7) \div 4$ m. The journey time will be given by distance ÷ speed = $0.5\pi \times 10^7$ m ÷ 20 m s⁻¹ = 8×10^5 s (about 9 days).

Now try this

1 Express each of the following in standard form.
   (a) mass of the Earth,
        5 972 000 000 000 000 000 000 000 kg　**(1 mark)**
   (b) mass of a proton,
        0.000 000 000 000 000 000 000 000 001 673 kg
        (1 mark)

2 Estimate the number of protons in your body.
   (5 marks)

3 Estimate the pressure exerted on the ground underneath an elephant's foot.　**(4 marks)**

You could estimate the elephant's mass by comparing its size with your own.

Experimental measurements

An experimental measurement is of limited use if we do not know the uncertainty associated with it.

Definitions

Accurate measurements are close to the accepted value (if there is one) for the quantity measured and they have a small range of uncertainties. They are made using correctly calibrated measuring devices.

- $m = 12.2\,kg \pm 0.2\,kg$ is more accurate than $m = 12.2\,kg \pm 0.4\,kg$.

The **precision** of a measurement is how exact it is. This is related to the number of significant figures that can be justified. It is also related to the finest scale division on the measuring device.

- $m = 12.20\,kg$ is quoted to higher precision than $m = 12.2\,kg$.

Errors and uncertainties

Making measurements is like aiming at the bull's-eye on a dartboard. There are two different types of error, **systematic** and **random errors**.

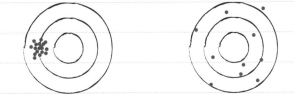

A systematic error shifts all values consistently. If you know about it, you can correct for it by subtracting the error from all readings. A zero error is an example of a systematic error.

A random error has no clear pattern to it. You can reduce the effect of random error by repeating the measurement and taking an average.

Worked example

A student measures the acceleration of free fall using this apparatus. Releasing a ball bearing from an electromagnet starts an electronic timer. When the ball bearing hits the contact pad the timer stops. The student repeats the experiment three times. Here are her results:

Distance between electromagnet and contact pad: 2.000 m

Diameter of ball bearing: 0.025 m

Time to fall (3 trials): 0.642 s, 0.646 s, 0.647 s

(a) Use these values to calculate a value for the acceleration of free fall, *g*. **(2 marks)**

Distance fallen = 2.000 – 0.025 m

Average time taken to fall

$= \dfrac{(0.642 + 0.646 + 0.647)}{3} = 0.645\,s$

$g = \dfrac{2s}{t^2} = \dfrac{(2 \times 1.975)}{0.645^2} = 9.49\,m\,s^{-2}$

ball release

Timer

contact pad

(b) The student initially forgot to subtract the diameter of the ball. What kind of error would this have introduced and what effect would it have had on the result? **(2 marks)**

This would introduce a systematic error. It would result in a larger value for g.

(c) What kind of error is reduced by repeating the measurements and taking an average? **(1 mark)**

This reduces random errors.

(d) The accepted value for *g* is 9.81 m s^{-2}. Suggest a reason for the difference between the student's value and the accepted value. **(3 marks)**

The student's value is too low. There is probably another systematic error that has not been taken into account. Maybe the timer started a short time after the ball was released so that all of the times are shorter.

Now try this

A student measures the time taken for a pendulum to complete 20 oscillations. He does this five times and records the results.

Trial 1 (s)	Trial 2 (s)	Trial 3 (s)	Trial 4 (s)	Trial 5 (s)
14.43	14.70	14.23	14.65	14.28

(a) Use the first trial to calculate the time for one oscillation. **(1 mark)**

(b) Use all of the data to calculate the average time for one oscillation. **(2 marks)**

(c) Explain why the result from (b) is more accurate than the result from (a) even though they are very similar. **(2 marks)**

Combining errors

When you calculate the value of a physical quantity from experimental data, the uncertainty in your result depends on the errors in each measurement, so you must understand how these errors combine.

Errors and uncertainties

Experimental 'errors' come about because of the limitations of the measuring instruments or methods used.

Uncertainties give the range of values within which a measured value lies.

$g = 9.81 \pm 0.02 \, \text{m s}^{-2}$. Here the uncertainty is $\pm 0.02 \, \text{m s}^{-2}$. The smaller the uncertainty, the more accurate the measured value.

Expressing uncertainty

For a measurement value x:

- absolute uncertainty Δx
- percentage (%) uncertainty $100 \times \dfrac{\Delta x}{x}$

You have measured the length of a wire to the nearest cm and obtained the value 1.25 m.

The absolute uncertainty $\Delta x = \pm 0.01 \, \text{m}$

The % uncertainty $100 \times \dfrac{\Delta x}{x} = \pm 0.8\%$

The masses are added together so the absolute uncertainty will be the sum of the absolute uncertainties for each mass.

Area is found by multiplying the lengths of the sides together, so the % uncertainties must be added.

Errors and uncertainties are estimated quantities so don't be afraid to round them off! Two significant figures for the final answer is usual, provided that you started with at least two.

The volume of the sphere can be calculated from $\frac{4}{3}\pi\left(\frac{d}{2}\right)^3$. Since d is cubed the % uncertainty in d must be multiplied by three.

Worked example

(a) A student measures two masses:
$m_1 = 10.25 \pm 0.05 \, \text{kg}$ and $m_2 = 8.50 \pm 0.05 \, \text{kg}$.
What is their combined mass M? **(2 marks)**

$M = (m_1 + m_2) \pm (\Delta m_1 + \Delta m_2)$
$= 18.75 \pm 0.10 \, \text{kg}$

(b) A sheet of paper has sides of lengths $30.0 \pm 0.2 \, \text{cm}$ and $20.0 \pm 0.2 \, \text{cm}$. What is its area A? **(3 marks)**

$A = 30.0 \times 20.0 = 600 \, \text{cm}^2$

% uncertainty in A
$= \left(\dfrac{0.2}{30} + \dfrac{0.2}{20}\right) \times 100 = 1.7\%$

Absolute uncertainty in A
$= 0.017 \times 600 = 10.2 \, \text{cm}^2$

$A = 600 \pm 10 \, \text{cm}^2$

(c) A student measures the diameter of a small ball bearing using a micrometer screw gauge and obtains a value of $5.00 \pm 0.01 \, \text{mm}$. What is the volume of the ball bearing? **(3 marks)**

Volume $= \frac{4}{3}\pi(2.50)^3 = 65.4 \, \text{mm}^3$

% uncertainty in volume
$= 3 \times \left(\dfrac{0.01}{5.00}\right) \times 100 = 0.6\%$

Absolute uncertainty in volume
$= 0.006 \times 65.4 = 0.39 \, \text{mm}^3$

Volume $= 65.4 \pm 0.4 \, \text{mm}^3$

Combining errors and uncertainties

When a value depends on several different measurements, you have to combine the errors to calculate the overall uncertainty in the value.

- ✓ If two measurements are added or subtracted: add the absolute uncertainties.
- ✓ If two measurements are multiplied or divided: add the % uncertainties.
- ✓ If a measurement is raised to the power n: multiply the % uncertainty by n.
- ✗ Never subtract uncertainties. The more measurements in the calculation, the larger the uncertainty becomes.

Now try this

A student carries out an experiment to measure the acceleration of free fall by dropping a small object through a measured height, s, and recording the time, t, it takes to reach the ground. His experimental results are: $s = 2.000 \pm 0.005 \, \text{m}$ and $t = 0.639 \pm 0.004 \, \text{s}$. Calculate the value for g, including the associated uncertainty, using the equation $g = \dfrac{2s}{t^2}$. **(4 marks)**

Graphs

Graphs are packed with information – it is important to understand the significance of gradients, intercepts and error bars.

Calculating the gradient of a force–extension graph

A student measures the extension, x, of a steel spring under different loading forces, F, with these results.

F (N)	2.0	4.0	6.0	8.0	10.0	12.0
x (m)	0.042 ± 0.005	0.078 ± 0.005	0.120 ± 0.005	0.161 ± 0.005	0.203 ± 0.005	0.237 ± 0.005

The absolute uncertainties in each measured value of x are used to plot error bars.

The student plots a graph of F versus x and uses the gradient to find the spring constant k. ———— $k = F \div x$

Often uncertainties in one quantity are much smaller than in the other one, so error bars need only be used for one of the variables.

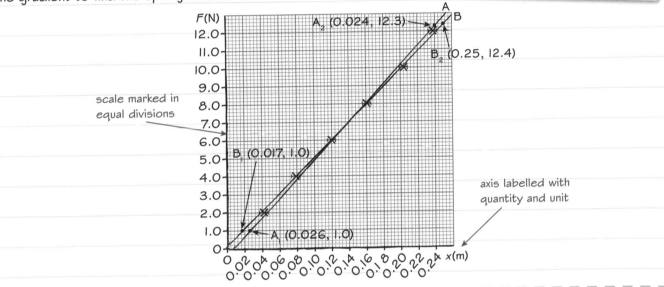

scale marked in equal divisions

axis labelled with quantity and unit

A_2 (0.024, 12.3)
B_2 (0.25, 12.4)
B_1 (0.017, 1.0)
A_1 (0.026, 1.0)

The two 'worst acceptable lines' intercept the F-axis above and below the origin, so the true line probably passes through the origin.

This suggests the spring obeys Hooke's law, $F = kx$.

The student chooses two points close to the ends of each line (A_1, B_1, A_2, B_2) to calculate the gradients.

$$\text{Gradient of A} = \left(\frac{12.3 - 1.0}{0.240 - 0.026}\right)$$
$$= 53\,\text{N m}^{-1}$$

$$\text{Gradient of B} = \left(\frac{12.4 - 1.0}{0.250 - 0.017}\right)$$
$$= 49\,\text{N m}^{-1}$$

When calculating a gradient from a graph always use points separated by at least half of the graph.

A gradient has units!

The units for the gradient are the units on the y-axis divided by the units on the x-axis.

From these data:

$$k = \left(\frac{53 + 49}{2}\right) = 51\,\text{N m}^{-1}$$

The uncertainty in k $\Delta k = \left(\frac{53 - 49}{2}\right) = 2\,\text{N m}^{-1}$

The spring constant $k = 51 \pm 2\,\text{N m}^{-1}$

Now try this

1 A student measures how the current through a resistor varies with the potential difference across it and plots a graph of V against I including error bars. The gradients of the two worst acceptable lines are $22.6\,\Omega$ and $22.0\,\Omega$. What is the measured value for the resistance? **(2 marks)**

2 Look at the graph above.

(a) What is the extension for a load of $5.0\,\text{N}$? **(2 marks)**

(b) What is the percentage uncertainty in the spring constant? **(1 mark)**

Scalars and vectors

Many physical quantities have a direction as well as a magnitude. You must take the direction into account when the quantities are combined.

Scalars and vectors

Scalars have magnitude but no direction.

Scalars include:

- mass
- speed
- temperature
- energy
- distance
- power.

> Distance is the magnitude of displacement, and speed is the magnitude of velocity.

Vectors have magnitude and direction.

Vectors include:

- displacement
- force
- velocity
- momentum.

Adding vectors

Vectors can be represented by arrows whose...

- length represents the magnitude of the vector
- direction represents the direction of the vector.

Find the sum (resultant, **R**) of two vectors **A** and **B** by putting them end to end.

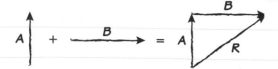

When two vectors are perpendicular we can calculate the magnitude of the resultant using Pythagoras' theorem. If the magnitudes of **A** and **B** above are 3 and 4 then the magnitude of **R** will be $5 = \sqrt{(3^2 + 4^2)}$.

Worked example

An aircraft is flying due north at $300\,\mathrm{m\,s^{-1}}$ and the wind is moving the air due east at $20\,\mathrm{m\,s^{-1}}$.

(a) What is the speed of the aircraft relative to the ground? **(3 marks)**

Use Pythagoras' theorem to find the magnitude of the resultant vector.

Drawing a vector diagram makes it much easier to solve the problem!

$v = \sqrt{(300^2 + 20^2)} = 301\,\mathrm{m\,s^{-1}}$

(b) In what direction does the aircraft travel? **(2 marks)**

The tangent of the angle to north is

$\tan\theta = \dfrac{20}{300} = 0.0667$. Therefore

$\theta = 3.8°$ E of N

Scalars and vectors on a running track

Most athletics tracks are ovals with a distance of 400 m per lap. The straights are the same length as the curves (100 m).

A child runs one lap in 80 s. Consider the distances and displacements of the child as he runs from A to B, C and D.

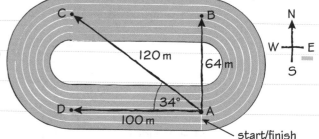

From A to ...	Distance	Displacement
B	100 m	64 m north (0°)
C	200 m	120 m 34°
D	300 m	100 m west
A	400 m	0 m

> See how taking the direction into account changes things!

Resolving vectors

Manchester is north and west of London. The displacement vector from London to Manchester can be resolved into two perpendicular components, one pointing north and one pointing west, with magnitudes $A\cos q$ and $A\sin q$.

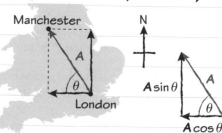

✓ Any vector can be resolved into components along perpendicular axes in this way.

Now try this

1 Look at the example of the 400 m runner above.
   (a) What is his average speed for one lap? **(1 mark)**
   (b) What is his average velocity for one lap? **(1 mark)**

2 A river is flowing due east at $1.2\,\mathrm{m\,s^{-1}}$. I can swim at a steady speed of $2.0\,\mathrm{m\,s^{-1}}$ relative to the water. I set off from the river bank swimming due north. What is my velocity vector relative to the bank? **(3 marks)**

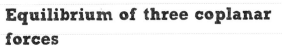
Vector triangles

Problems involving three vectors can often be solved by drawing a triangle and finding the lengths of its sides or the sizes of its angles.

Finding the resultant of two coplanar vectors

When two vectors are added the resultant can be found from the third side of the vector triangle.

The object right has two forces acting on it.
The resultant is found by solving the right-angled triangle.

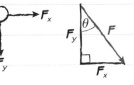

Vector $\mathbf{F} = \mathbf{F}_x + \mathbf{F}_y$

Magnitude $F^2 = F_x^2 + F_y^2$ (Pythagoras)

so $F = \sqrt{(F_x^2 + F_y^2)}$

$\tan\theta = \dfrac{F_x}{F_y}$

Maths skills — Relative velocity problems (addition of vectors)

The velocity vector relative to the bank, $\mathbf{v}$, of a fish swimming in a moving river is the sum of the fish's velocity relative to the river, $\mathbf{v}_F$, and the velocity of the river relative to the bank, $\mathbf{v}_R$.

Solve problems like this using a velocity vector triangle.

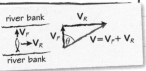

Equilibrium of three coplanar forces

If all the forces acting on a body are in equilibrium then their resultant is zero. This means that when drawn end to end they must form a closed polygon. If there are just three forces this will form a **triangle of forces**. An unknown force can be found by solving the triangle.

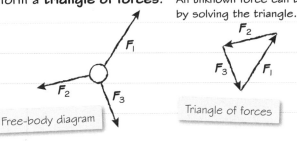

Free-body diagram

Triangle of forces

✓ If it is a right-angled triangle then Pythagoras' theorem can be used to solve for the magnitude or angle of the unknown vector.

Relative velocity problems (subtraction of vectors)

Imagine two aircraft, A and B, travelling in different directions. The relative velocity, $\mathbf{v}_{AB}$, of A as it appears to B is found by subtracting the velocity of B relative to the air, $\mathbf{v}_B$, from the velocity of A relative to the air, $\mathbf{v}_A$.

To subtract a vector, simply add a vector of the same magnitude pointing in the opposite direction.

$$\mathbf{v}_{AB} = \mathbf{v}_A - \mathbf{v}_B$$

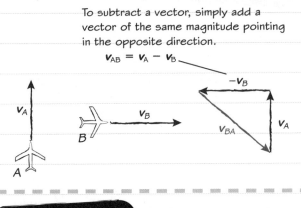

Worked example

The diagram shows a car parked on a hill. The hill is at an angle of 10° to the horizontal. Three forces act on the car: its weight of 12 kN, a normal contact force N at 90° to the road and a frictional force G parallel to the road. Use a triangle of forces to find the magnitudes of G and N. **(2 marks)**

The forces are in equilibrium so they form a triangle. G and N are perpendicular so it is a right-angled triangle.

$N = W\cos10° = 12\,000\cos10°$
$\quad = 11\,800\,\text{N}$

$G = W\sin10° = 12\,000\sin10°$
$\quad = 2100\,\text{N}$

G could also be found using W and N and applying Pythagoras' theorem, or both N and G could be found by scale drawing.

Now try this

1 Look at the diagram of the car on the hill. Find the magnitudes of N and G if the road makes an angle of 20° to the horizontal. **(2 marks)**

2 A ferry boat sails between two jetties on opposite sides of a wide river that flows due east. The boat has a speed relative to the water of $5.0\,\text{m s}^{-1}$ and the river is flowing at $3.0\,\text{m s}^{-1}$. In what direction must the boat set sail from the southern side in order to travel directly to the jetty on the northern side? **(3 marks)**

Resolving vectors

Any vector can be resolved into two perpendicular components.

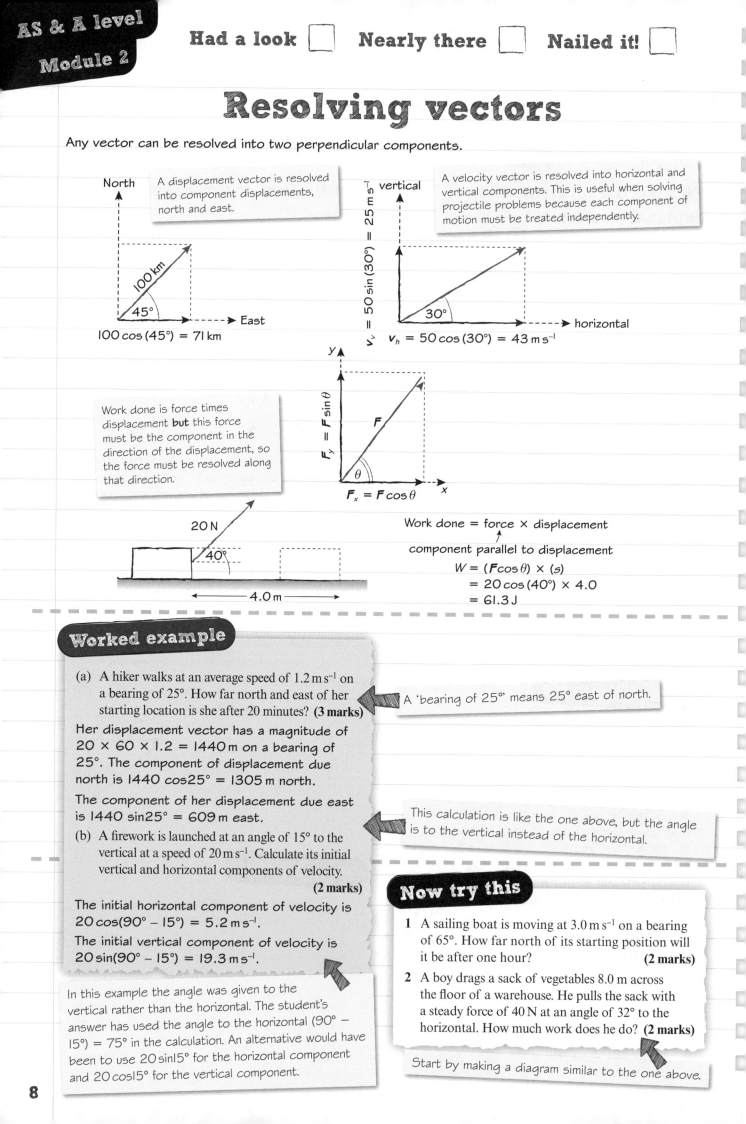

North

A displacement vector is resolved into component displacements, north and east.

$100 \cos(45°) = 71$ km

A velocity vector is resolved into horizontal and vertical components. This is useful when solving projectile problems because each component of motion must be treated independently.

$v_v = 50 \sin(30°) = 25$ m s^{-1}

$v_h = 50 \cos(30°) = 43$ m s^{-1}

Work done is force times displacement **but** this force must be the component in the direction of the displacement, so the force must be resolved along that direction.

$F_y = F \sin\theta$

$F_x = F \cos\theta$

Work done = force × displacement

↑ component parallel to displacement

$W = (F\cos\theta) \times (s)$
$= 20\cos(40°) \times 4.0$
$= 61.3$ J

Worked example

(a) A hiker walks at an average speed of 1.2 m s^{-1} on a bearing of 25°. How far north and east of her starting location is she after 20 minutes? **(3 marks)**

Her displacement vector has a magnitude of $20 \times 60 \times 1.2 = 1440$ m on a bearing of 25°. The component of displacement due north is $1440 \cos25° = 1305$ m north.

The component of her displacement due east is $1440 \sin25° = 609$ m east.

(b) A firework is launched at an angle of 15° to the vertical at a speed of 20 m s^{-1}. Calculate its initial vertical and horizontal components of velocity.

(2 marks)

The initial horizontal component of velocity is $20\cos(90° - 15°) = 5.2$ m s^{-1}.

The initial vertical component of velocity is $20\sin(90° - 15°) = 19.3$ m s^{-1}.

A 'bearing of 25°' means 25° east of north.

This calculation is like the one above, but the angle is to the vertical instead of the horizontal.

In this example the angle was given to the vertical rather than the horizontal. The student's answer has used the angle to the horizontal $(90° - 15°) = 75°$ in the calculation. An alternative would have been to use $20 \sin15°$ for the horizontal component and $20 \cos15°$ for the vertical component.

Now try this

1 A sailing boat is moving at 3.0 m s^{-1} on a bearing of 65°. How far north of its starting position will it be after one hour? **(2 marks)**

2 A boy drags a sack of vegetables 8.0 m across the floor of a warehouse. He pulls the sack with a steady force of 40 N at an angle of 32° to the horizontal. How much work does he do? **(2 marks)**

Start by making a diagram similar to the one above.

Describing motion

To understand and analyse motion you need a clear idea of how displacement, velocity and acceleration are defined and how they are linked to one another.

Definitions

Velocity is a vector – it is speed in a particular direction.

	Symbol	Unit
Displacement	s	m
Velocity	u or v	$m\,s^{-1}$
Acceleration	a	$m\,s^{-2}$

u is an initial velocity
v is a final velocity

Symbol represents a physical quantity. —— $t = 25\,s$ —— Unit for measure of a physical quantity.

Rates of change

Velocity is the rate of change of displacement.

$$v = \frac{\text{change in displacement}}{\text{time taken}} = \frac{\Delta s}{\Delta t}$$

Acceleration is the rate of change of velocity.

$$a = \frac{\text{change in velocity}}{\text{time taken}} = \frac{\Delta v}{\Delta t}$$

or

$$a = \frac{(v - u)}{t}$$

The values of velocity and acceleration are average values over the time Δt. To get an instantaneous value the time must approach zero.

Δ

'Δ' means 'change in', so Δv is a change in velocity.

For example, if a car accelerates from $u = 20\,m\,s^{-1}$ to $v = 25\,m\,s^{-1}$ to overtake a truck, its change in velocity Δv is:

$$\Delta v = v - u = 25 - 20 = 5\,m\,s^{-1}$$

Worked example

The table below gives the displacement, s, of a sprinter for each second of a 100 m race.

t (s)	0.0	1.0	2.0	3.0	4.0	5.0	6.0	7.0	8.0	9.0	10.0
s (m)	0.0	2.0	8.0	18.0	30.0	42.0	54.0	66.0	78.0	90.0	102.0

(a) Calculate the average velocity during the first second.   **(2 marks)**

$\Delta s = 2.0 - 0.0 = 2.0\,m$ and $\Delta t = 1.0\,s$

$$v = \frac{\Delta s}{\Delta t} = \frac{2.0}{1.0} = 2.0\,m\,s^{-1}$$

The known values of Δs and Δt have been listed before the calculation is carried out.

(b) Calculate the average velocity between $t = 1.0\,s$ and $t = 2.0\,s$.   **(2 marks)**

$\Delta s = 8.0 - 2.0 = 6.0\,m$ and $\Delta t = 1.0\,s$.

$$v = \frac{\Delta s}{\Delta t} = \frac{6.0}{1.0} = 6.0\,m\,s^{-1}.$$

(c) Calculate the average acceleration between $t = 1\,s$ and $t = 2\,s$.   **(3 marks)**

$\Delta v = 6.0 - 2.0 = 4.0\,m\,s^{-1}$ and $\Delta t = 1.0\,s$.

$$a = \frac{\Delta v}{\Delta t} = \frac{4.0}{1.0} = 4.0\,m\,s^{-2}.$$

Now try this

Look at the table in the worked example above.

(a) Calculate the average velocity of the sprinter during the 10 s recorded in the table.   **(2 mark)**

(b) Calculate the average velocity of the sprinter between $t = 5.0\,s$ and $t = 10.0\,s$.   **(2 mark)**

(c) What was the acceleration of the sprinter between $t = 5.0\,s$ and $t = 10\,s$? Explain your answer.   **(2 marks)**

(d) Calculate the time taken for the sprinter to complete this 100 m race.   **(3 marks)**

Graphs of motion

Make sure you are confident about the different types of information you can get from graphs of motion.

Representing motion

Here are three graphs of motion for an object moving with constant acceleration.

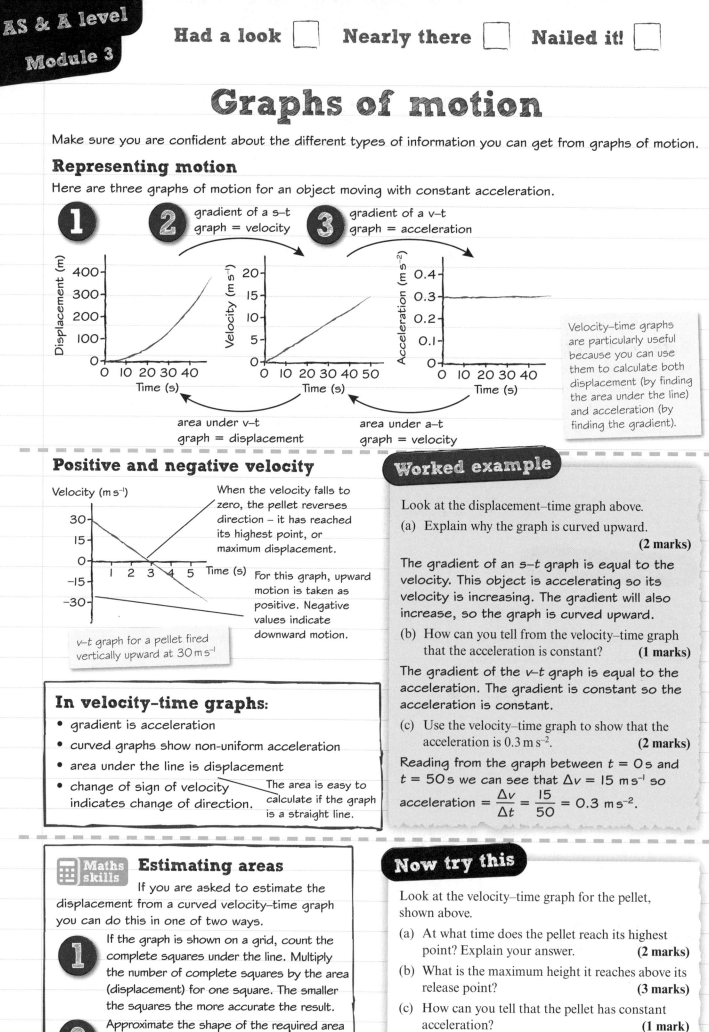

1

2 gradient of a s–t graph = velocity

3 gradient of a v–t graph = acceleration

area under v–t graph = displacement

area under a–t graph = velocity

Velocity–time graphs are particularly useful because you can use them to calculate both displacement (by finding the area under the line) and acceleration (by finding the gradient).

Positive and negative velocity

When the velocity falls to zero, the pellet reverses direction – it has reached its highest point, or maximum displacement.

For this graph, upward motion is taken as positive. Negative values indicate downward motion.

v–t graph for a pellet fired vertically upward at 30 m s⁻¹

In velocity–time graphs:

* gradient is acceleration
* curved graphs show non-uniform acceleration
* area under the line is displacement
* change of sign of velocity indicates change of direction.

The area is easy to calculate if the graph is a straight line.

Worked example

Look at the displacement–time graph above.

(a) Explain why the graph is curved upward.
(2 marks)

The gradient of an s–t graph is equal to the velocity. This object is accelerating so its velocity is increasing. The gradient will also increase, so the graph is curved upward.

(b) How can you tell from the velocity–time graph that the acceleration is constant? **(1 marks)**

The gradient of the v–t graph is equal to the acceleration. The gradient is constant so the acceleration is constant.

(c) Use the velocity–time graph to show that the acceleration is $0.3 \, \text{m s}^{-2}$. **(2 marks)**

Reading from the graph between $t = 0$ s and $t = 50$ s we can see that $\Delta v = 15 \, \text{m s}^{-1}$ so

acceleration $= \dfrac{\Delta v}{\Delta t} = \dfrac{15}{50} = 0.3 \, \text{m s}^{-2}$.

Maths skills Estimating areas

If you are asked to estimate the displacement from a curved velocity–time graph you can do this in one of two ways.

1 If the graph is shown on a grid, count the complete squares under the line. Multiply the number of complete squares by the area (displacement) for one square. The smaller the squares the more accurate the result.

2 Approximate the shape of the required area using triangles or rectangles and use the area of these to estimate the displacement.

Now try this

Look at the velocity–time graph for the pellet, shown above.

(a) At what time does the pellet reach its highest point? Explain your answer. **(2 marks)**

(b) What is the maximum height it reaches above its release point? **(3 marks)**

(c) How can you tell that the pellet has constant acceleration? **(1 mark)**

(d) Calculate the acceleration of the pellet and state the direction of this acceleration. **(3 marks)**

SUVAT equations of motion

Four important equations link displacement, velocity and acceleration during periods of constant acceleration.

Definitions

s = displacement (m)

u = initial velocity (m s⁻¹)

v = final velocity (m s⁻¹)

a = acceleration (m s⁻²)

t = time (s)

Key points about SUVAT equations:

- You can only apply the SUVAT equations for periods with **constant** acceleration.
- Negative acceleration is deceleration.
- If a question says that something 'starts from rest' it is giving you the initial velocity: $u = 0\,\text{m s}^{-1}$.
- If a question says that something comes to a stop, it is telling you that $v = 0\,\text{m s}^{-1}$.
- If a question says that an object is dropped then $a = g = 9.81\,\text{m s}^{-2}$ (on Earth).

To solve problems involving constant acceleration:

✓ Read the question carefully!

✓ List the values given in the question.

✓ Identify the quantity that needs to be calculated.

✓ Identify the SUVAT equation which contains the unknown **and** three known quantities.

✓ Rearrange and solve this equation.

SUVAT equations

Remember these four equations:

1 $v = u + at$

2 $s = \frac{1}{2}(u + v)t$

3 $s = ut + \frac{1}{2}at^2$

4 $v^2 = u^2 + 2as$

Maths skills Rearranging is like solving an equation. You have to do the same thing to both sides:

$v = u + at$ $(- u)$

$v - u = at$ $(\div t)$

$\dfrac{(v - u)}{t} = a$

Worked example

A cheetah can accelerate from rest to $28\,\text{m s}^{-1}$ in just 3.5 s!

(a) Calculate the cheetah's acceleration (assume it is constant). **(2 marks)**

$u = 0\,\text{m s}^{-1}$ $v = 28\,\text{m s}^{-1}$ $t = 3.5\,\text{s}$ $a = ?$

$v = u + at$

$a = \dfrac{(v - u)}{t} = \dfrac{28}{3.5} = \underline{8.0\ \text{m s}^{-2}}$

Write out the SUVAT values that you are given in the question. You want to find acceleration, so write $a = ?$. You can use equation **1**, $v = u + at$. Write out the equation then rearrange it to make a the subject.

(b) Calculate how far it moves while it accelerates. **(2 marks)**

$u = 0\,\text{m s}^{-1}$ $v = 28\,\text{m s}^{-1}$ $t = 3.5\,\text{s}$ $s = ?$

$s = \frac{1}{2}(u + v)t = \frac{1}{2}(0 + 28) \times 3.5 = \underline{49\,\text{m}}$

You could use equation **2** or **3**. Use equation **2** because you don't need to use your answer from part (a).

Now try this

1 A sports car can accelerate from $0\,\text{m s}^{-1}$ to $50\,\text{m s}^{-1}$ in 10 s.

  (a) Calculate the acceleration of the sports car. **(2 marks)**

  (b) Calculate the distance travelled by the sports car as it accelerates from $0\,\text{m s}^{-1}$ to $50\,\text{m s}^{-1}$. **(2 marks)**

2 A cyclist is travelling at $15.0\,\text{m s}^{-1}$ along a straight road and begins to go down a hill. On the way downhill he has a constant acceleration of $0.50\,\text{m s}^{-2}$. At the bottom of the hill he is travelling at $25\,\text{m s}^{-1}$. Calculate the distance he travelled down the hill. **(3 marks)**

Acceleration of free fall

In the absence of friction, all objects in the same gravitational field fall at the same rate.

Key points about free fall

- An object is said to be in **free fall** if the only force acting on it is gravity.

- In a uniform gravitational field the acceleration due to free fall is constant.

- Near the surface of the Earth the acceleration of free fall is about $9.81\,\mathrm{m\,s^{-2}}$.

- Near the surface of the Moon the acceleration of free fall is about $1.6\,\mathrm{m\,s^{-2}}$.

- Objects of different mass have the same acceleration in free fall.

- The symbol used to represent the acceleration of free fall is g.

Satellites orbiting the Earth are in free fall, as are astronauts inside orbiting spacecraft.

Apollo astronauts dropped a feather and hammer from the same height on the Moon. They fell at the same rate, as there was no friction from air, and reached the ground together.

The acceleration of free fall is equal to the gravitational field strength, so g is used for both quantities and the units $\mathrm{m\,s^{-2}}$ are equivalent to $\mathrm{N\,kg^{-1}}$.

Practical skills

Measuring g in the laboratory

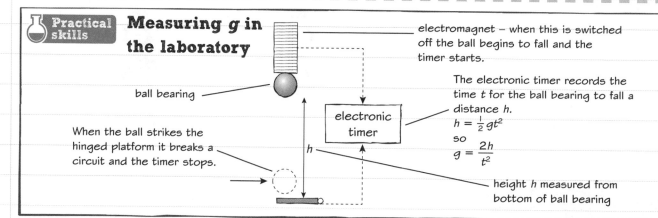

electromagnet – when this is switched off the ball begins to fall and the timer starts.

ball bearing

The electronic timer records the time t for the ball bearing to fall a distance h.

$h = \frac{1}{2}gt^2$

so

$g = \frac{2h}{t^2}$

When the ball strikes the hinged platform it breaks a circuit and the timer stops.

electronic timer

h

height h measured from bottom of ball bearing

Worked example

A student uses an electronic timer to measure the acceleration of free fall. She measures the height of the drop to be 50.0 cm and makes three successive measurements of the time to fall: 322 ms, 325 ms and 327 ms. Calculate the acceleration of free fall g in this experiment. **(3 marks)**

Height $h = 0.500\,\mathrm{m}$

The average time t to fall is

$t = \dfrac{(323 + 325 + 327)}{3} = 325\,\mathrm{ms} = 0.325\,\mathrm{s}$

$g = \dfrac{2h}{t^2} = \underline{9.47\,\mathrm{m\,s^{-2}}}$

Repeating an experiment and taking an average reduces the effect of random errors.

If you have a large number of experimental values for two variables you can plot a graph to find an average value. Here you would plot measurements of $2h$ against t^2 and find g from the gradient.

Now try this

The student decides to measure g using a larger drop height of 1.20 m. She takes five measurements of the time to fall. These are: 497 ms, 500 ms, 496 ms, 476 ms and 495 ms.

(a) Which result is anomalous? What should she do about this result? **(2 marks)**

(b) Calculate the average time to fall. **(1 mark)**

(c) Calculate the acceleration of free fall from the data. **(2 marks)**

Vehicle stopping distances

The total stopping distance of a car is the sum of the thinking distance and the braking distance.

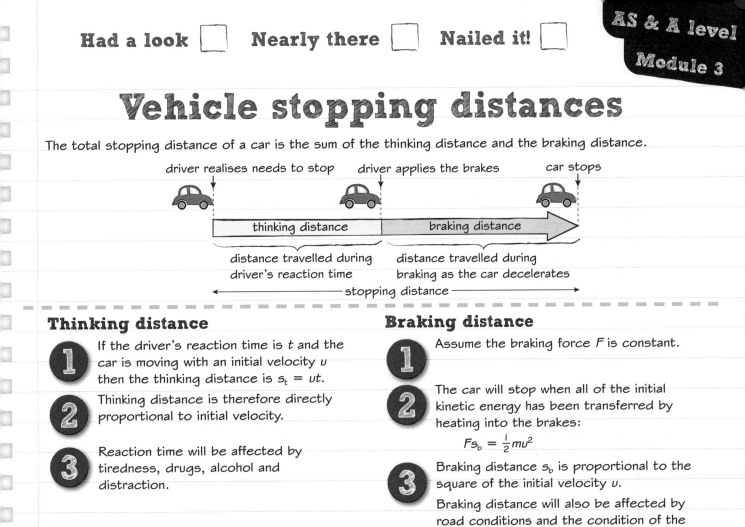

Thinking distance

1 If the driver's reaction time is t and the car is moving with an initial velocity u then the thinking distance is $s_t = ut$.

2 Thinking distance is therefore directly proportional to initial velocity.

3 Reaction time will be affected by tiredness, drugs, alcohol and distraction.

Braking distance

1 Assume the braking force F is constant.

2 The car will stop when all of the initial kinetic energy has been transferred by heating into the brakes:

$$Fs_b = \tfrac{1}{2}mu^2$$

3 Braking distance s_b is proportional to the square of the initial velocity u.

Braking distance will also be affected by road conditions and the condition of the car (tyres/brakes).

Variation of stopping distance with initial velocity

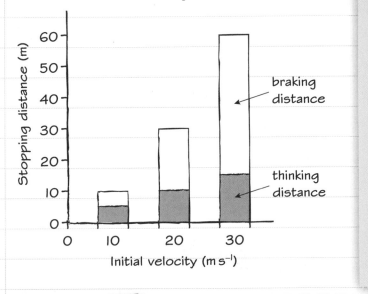

Worked example

Look at the bar chart of typical stopping distances.

(a) Explain why doubling the initial velocity doubles the thinking distance but more than doubles the braking distance. **(2 marks)**

Thinking distance is directly proportional to initial velocity but braking distance is proportional to velocity squared.

(b) Calculate the stopping distance for the same car travelling at an initial velocity of $40\,\mathrm{m\,s^{-1}}$ and state your assumptions. **(3 marks)**

The thinking distance will be twice the value at $20\,\mathrm{m\,s^{-1}}$ (20 m) and the braking distance will be 2^2 times the value at $20\,\mathrm{m\,s^{-1}}$ (80 m). The stopping distance will be (20 + 80) = 100 m. We have assumed the same constant braking force and the same reaction time.

Now try this

(a) Use the data in the bar chart above to calculate the thinking distance for the same car travelling at an initial velocity of $25\,\mathrm{m\,s^{-1}}$ (about 55 miles per hour). **(2 marks)**

(b) Use the data in the bar chart above to calculate the braking distance for the same car travelling at an initial velocity of $25\,\mathrm{m\,s^{-1}}$. **(2 marks)**

(c) Use your values from (a) and (b) to calculate the total stopping distance for the car for an initial velocity of $25\,\mathrm{m\,s^{-1}}$ and state any assumptions you have made. **(2 marks)**

Projectile motion

The horizontal and vertical components of motion are **independent** of one another.

Using SUVAT equations

SUVAT equations can be used to solve problems of motion of projectiles that move in two dimensions.

If one object is dropped at the same time as another is projected horizontally from the same height, both will have the same vertical motion and hit the ground at the same time.

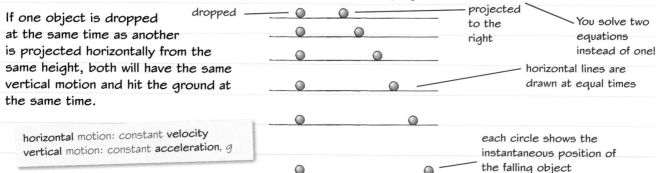

dropped — projected to the right

You solve two equations instead of one!

horizontal lines are drawn at equal times

each circle shows the instantaneous position of the falling object

horizontal motion: constant **velocity**
vertical motion: constant **acceleration**, g

Horizontal projectile

If an object is projected horizontally in a gravitational field of strength, g:

- vertical distance is given by $s_v = \frac{1}{2}gt^2$
- horizontal distance is given by $s_h = ut$ where u is the initial horizontal velocity.

Notice how the student has treated each part of the motion (horizontal and vertical) independently and used subscripts to distinguish the two parts of the motion.

Range of a projectile

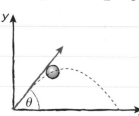

If an object is projected at initial speed u at an angle θ to the horizontal:

- The initial horizontal velocity is $u\cos\theta$.
- The initial vertical velocity is $u\sin\theta$.

To analyse the motion you must treat the horizontal and vertical motions as separate linear motions.

A bowler in a cricket match projects the cricket ball horizontally at $30\,\text{m s}^{-1}$ from a height of $2.0\,\text{m}$ above the ground.

(a) How long does the ball take to reach the ground?
(3 marks)

For vertical motion, we know that

$s_v = 2.0\,\text{m}, g = 9.81\,\text{m s}^{-2}$

$s_v = \frac{1}{2}gt^2$ so $t = \sqrt{\left(\dfrac{2s_v}{g}\right)} = \sqrt{\left(\dfrac{4.0}{9.81}\right)} = 0.64\,\text{s}$

(b) How far does the ball travel horizontally before it hits the ground?
(1 mark)

For horizontal motion, we know u_h and we have just calculated t.

$s_h = u_h t = 30 \times 0.64 = 19\,\text{m}$

Maths skills Velocity is a vector. These are the horizontal and vertical components of the velocity vector. See page 8 for a reminder about resolving vectors.

To calculate horizontal range:

1 Time for the vertical velocity to fall to zero (top of the flight) is $t = \dfrac{(u\sin\theta)}{g}$

2 Time of flight is $2t$ (it goes up and then comes down)

3 Range $= u\cos\theta \times 2t = \dfrac{(2u^2\sin\theta\cos\theta)}{g}$
This has a maximum value when $\theta = 45°$.

1 A low-flying aircraft releases an aid package while it is flying horizontally at $50\,\text{m s}^{-1}$ at an altitude of $40\,\text{m}$. How far before the drop zone should it release the load? (Ignore the effects of friction.)
(4 marks)

2 A cannonball is fired at an angle of $30°$ to the horizontal. The initial velocity of the cannonball is $60\,\text{m s}^{-1}$. Calculate its range and say why, in practice, the actual range is likely to be less than this.
(4 marks)

Types of force

There are many different types of force but they are all defined by how they affect motion.

Force, mass and acceleration

If the net (resultant) force acting on something is zero it will stay at rest or continue moving with constant velocity.

The forces balance, that is, cancel each other out.

If the force is unbalanced the velocity of the object will change – it will accelerate.

This can mean speeding up, slowing down or changing direction.

The size of the force is defined by the equation:

$$F = ma$$

where F is the net force on the object.

Units of force

The newton (N) is the S.I. unit of force.

A net force of 1 N will cause a mass of 1 kg to accelerate at 1 m s⁻².

Types of force

The forces you know can be split into:

1 Gravitational – weight is a gravitational force that acts on mass. $W = mg$, where g is the gravitational field strength, which is the same as the acceleration due to gravity.

about 9.81 N kg⁻¹ or 9.81 m s⁻² near the Earth's surface

2 Electromagnetic – electromagnetic forces act on charges. Ordinary matter is made of atoms which contain charged particles, so many common forces, such as the friction or the contact force when two objects touch, arise from electromagnetic interactions.

3 Nuclear – two types of nuclear force act on particles in nuclei.

Representing forces

Forces are **vectors** so they can be drawn as arrows.
- The arrow direction shows the direction of the force.
- The length of the arrow shows the magnitude of the force.
- The diagram shows four forces acting on a car moving at constant velocity along a straight horizontal road.

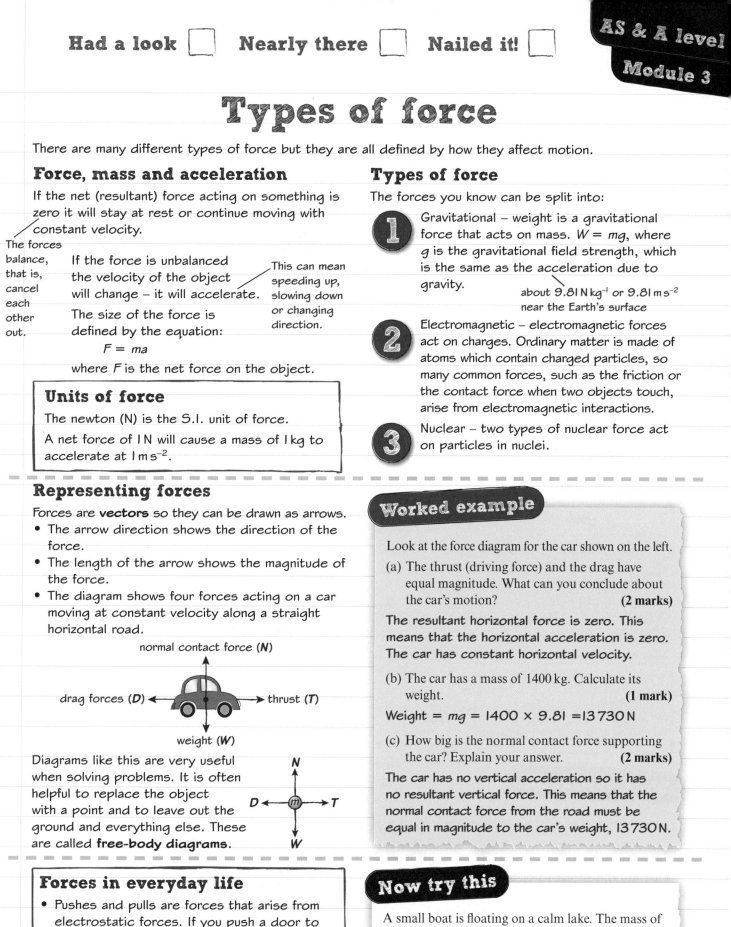

normal contact force (**N**)

drag forces (**D**) ← → thrust (**T**)

weight (**W**)

Diagrams like this are very useful when solving problems. It is often helpful to replace the object with a point and to leave out the ground and everything else. These are called **free-body diagrams**.

N
D ← (m) → T
W

Look at the force diagram for the car shown on the left.

(a) The thrust (driving force) and the drag have equal magnitude. What can you conclude about the car's motion?   **(2 marks)**

The resultant horizontal force is zero. This means that the horizontal acceleration is zero. The car has constant horizontal velocity.

(b) The car has a mass of 1400 kg. Calculate its weight.   **(1 mark)**

Weight = mg = 1400 × 9.81 = 13 730 N

(c) How big is the normal contact force supporting the car? Explain your answer.   **(2 marks)**

The car has no vertical acceleration so it has no resultant vertical force. This means that the normal contact force from the road must be equal in magnitude to the car's weight, 13 730 N.

Forces in everyday life

- Pushes and pulls are forces that arise from electrostatic forces. If you push a door to open it the electrons in atoms on the surface of your hand repel electrons in atoms on the surface of the door.
- Tension in a stretched spring also arises from the electrostatic forces between atoms inside the stretched material. As atoms are pulled further apart their attraction increases (unless the bond breaks).

A small boat is floating on a calm lake. The mass of the boat is 250 kg.

(a) Calculate the weight of the boat.   **(1 mark)**

(b) State the value of the upthrust on the boat and explain why it must have this value.   **(2 marks)**

(c) A child of mass 50 kg steps into the boat and his father on the bank applies a forward force of 60 N to the boat. Calculate the initial acceleration of the boat and child.   **(2 marks)**

Drag

Drag forces oppose motion and often vary with velocity.

Terminal velocity

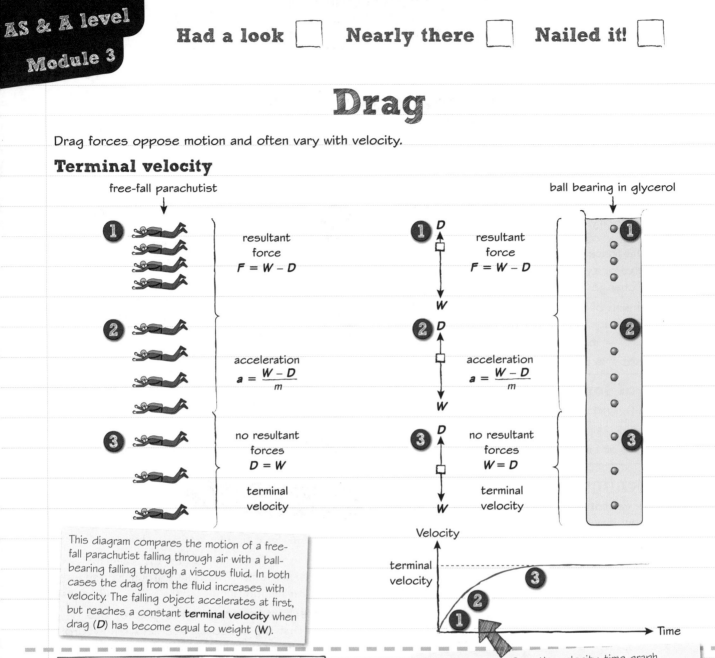

free-fall parachutist

ball bearing in glycerol

1 resultant force $F = W - D$

2 acceleration $a = \dfrac{W - D}{m}$

3 no resultant forces $D = W$

terminal velocity

1 resultant force $F = W - D$

2 acceleration $a = \dfrac{W - D}{m}$

3 no resultant forces $W = D$

terminal velocity

This diagram compares the motion of a free-fall parachutist falling through air with a ball-bearing falling through a viscous fluid. In both cases the drag from the fluid increases with velocity. The falling object accelerates at first, but reaches a constant **terminal velocity** when drag (D) has become equal to weight (W).

Positions 1, 2 and 3 on the velocity–time graph correspond to the labelled positions on the diagrams above. Note that the graph is the same shape for both objects (but will have different scales). This is because the underlying physics is the same.

Drag is affected by:

* velocity
* air density (or density of the medium)
* shape and size of the object.

Practical skills Measuring terminal velocity: when the ball bearing in glycerol reaches terminal velocity at **3** a stopwatch is used to time its fall through a measured distance.

Worked example

A stone is released from just beneath the surface of a deep lake. State and explain how its velocity changes as it falls to the bottom of the lake. **(5 marks)**

Initially the stone has a low velocity and a large downward acceleration. The drag is much smaller than its weight because it is moving slowly. As it falls its velocity increases and the acceleration decreases. This is because the drag increases with velocity so the resultant force decreases. Eventually (if it has not hit the bottom of the lake) it will reach a constant terminal velocity. This is because the drag has become equal to the weight so the resultant force is zero.

Now try this

(a) Explain why a feather reaches a terminal velocity when dropped on the Earth but not when dropped on the Moon. **(2 marks)**

(b) Calculate the downward acceleration of a falling ball of mass 200 g when the drag force on it is 1.0 N. **(3 marks)**

(c) Explain why a car has a maximum speed even though the engine continues to provide a forward force. **(2 marks)**

Centre of mass and centre of gravity

It is easier to solve problems if we treat a large object as if it were a point mass.

Centre of mass

A solid object can be treated as if all of its mass is concentrated at a single point, the **centre of mass**.

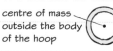

centre of mass at the geometric centre of a sphere of uniform density

centre of mass outside the body of the hoop

Finding centre of gravity for a plane (laminar) object

1 Suspend the object from a point and use a plumb line to mark a vertical line on the object.

2 Repeat using a different point of suspension.

3 The centre of gravity is at the point where the two lines intersect.

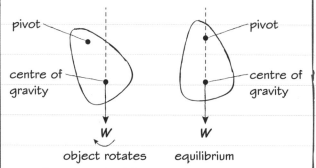

pivot

centre of gravity

W

object rotates

pivot

centre of gravity

W

equilibrium

Centre of gravity

When gravity acts on an extended object it pulls on every atom separately. The sum of all of these forces is the weight of the object.

The point through which the weight of the object acts is the **centre of gravity** of the object.

If the object is in a uniform gravitational field (e.g. on Earth) then the centre of gravity and centre of mass are in the same place.

When an object is suspended from a point it will come to equilibrium with its centre of gravity directly beneath the point of suspension.

centre of gravity

For a gymnast/tightrope walker to balance, the centre of gravity must be directly above the area of contact with the bar/wire.

A student investigated the minimum angle at which a plastic bottle would topple over when tilted away from the vertical. She thought that the more water she added the more stable it would become. Here are her results.

	Empty	$\frac{1}{4}$ full	$\frac{1}{2}$ full	$\frac{3}{4}$ full	full
Angle	15°	25°	20°	18°	15°

(a) Explain how these results contradict her expectation. **(2 marks)**

If she had been correct the angle needed to topple the bottle would have increased to a maximum when full. In fact the maximum angle occurs when it is only $\frac{1}{4}$ full.

(b) Use the idea of centre of gravity to explain why the bottle was easiest to topple when it was empty or full. **(2 marks)**

In both cases the centre of mass would be about half way up the bottle. When the bottle is only partly full the centre of mass of the water is lower, below half way, so it has to be tilted through a larger angle before the centre of gravity passes outside the base and the bottle topples.

1 Explain why the bottle in the example above topples over when it is tilted through a large enough angle. **(3 marks)**

2 Explain why the bottle has maximum stability when it is about $\frac{1}{4}$ full. **(2 marks)**

Moments, couples and torques

The turning effect of one or more forces can change the way a body rotates.

Moment of a force

Moment = force × perpendicular distance from pivot
The S.I. unit for a moment is the N m.

moment = **F**d

pivot

90°

F

d

Note how *d* is the perpendicular distance in the example above. It is different from the distance between the pivot and the point of action of the force.

moment = m**g**d

pivot

d

mg

Couples

When two parallel forces of equal magnitude act in opposite directions on the same body they produce a couple. The moment or **torque** of a couple is equal to the sum of the moments caused by each individual force.

F
90°
d
90°
F

This turns out to be the same value about any point in the body and is given by:

Torque (moment of a couple)
= force × perpendicular distance between forces

Beware newton metres!

S.I. units for moments and for energy are N m, but these are not equivalent. For moments the force and distance are perpendicular to one another. For energy they are parallel.

To solve an equilibrium problem:

1 Choose a suitable point to act as the pivot.

2 Calculate the sum of clockwise moments.

3 Calculate the sum of anticlockwise moments.

4 Equate the total clockwise and anticlockwise moments.

The principle of moments

For a body in equilibrium, the sum of clockwise moments is equal to the sum of anticlockwise moments about any point.

If all the forces that act upon an object are balanced, then the object does not rotate and is said to be in a state of equilibrium.

Worked example

The system below is balanced.
Find the mass of *X*. **(4 marks)**

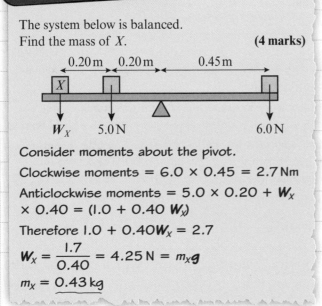

0.20 m   0.20 m   0.45 m

X

W$_X$ 5.0 N   6.0 N

Consider moments about the pivot.

Clockwise moments = 6.0 × 0.45 = 2.7 N m

Anticlockwise moments = 5.0 × 0.20 + *W*$_X$ × 0.40 = (1.0 + 0.40 *W*$_X$)

Therefore 1.0 + 0.40*W*$_X$ = 2.7

$W_X = \dfrac{1.7}{0.40} = 4.25$ N = $m_X g$

$m_X = 0.43$ kg

Now try this

The gymnast in the diagram above has a mass of 60 kg and his body is at 30° to the horizontal. Calculate the resultant moment on his body if the distance from his grip on the bar to his centre of gravity is 1.2 m. **(3 marks)**

Remember that you need the perpendicular distance from the pivot. The diagram shows a right-angled triangle. The question tells you the angle and the length of the hypotenuse, so you can calculate *d*.

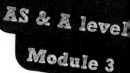
Equilibrium

An object is in equilibrium when the forces and moments acting on it are all balanced, so there is no resultant.

Forces in equilibrium

When forces are in equilibrium a body will have zero acceleration. Forces are vectors, so an object can only be in equilibrium if all components of forces are balanced.

It may be at rest, but it could be moving with constant velocity.

To solve equilibrium problems you often have to consider horizontal and vertical components of force separately. It helps to draw a free-body diagram showing all forces acting on the body first.

Triangle of forces

When three coplanar forces act on a body and they are in equilibrium, a force diagram (with force vectors drawn end to end) forms a closed triangle of forces. This fact gives us an alternative method for solving equilibrium problems involving three coplanar forces.

In the example below the pendulum bob has been pulled sideways and is being held in equilibrium. Therefore the resultant force acting on it is zero.

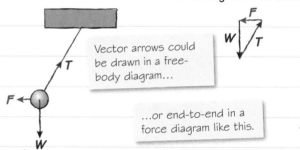

Vector arrows could be drawn in a free-body diagram...

...or end-to-end in a force diagram like this.

If two of the forces acting are known, then the third force can be found by solving the triangle.

Solving equilibrium problems for moments and forces

Forces

1 Resolve forces along two perpendicular directions (often vertical and horizontal).

2 For each direction equate total forces acting in opposite directions.

Choosing a point through which one or more forces act will simplify the problem!

Moments

1 Select a suitable pivot

2 Use the principle of moments to equate clockwise and anticlockwise moments.

Worked example

A boy holds a heavy drain cover open at an angle of 30° to the horizontal as shown in the diagram below. The weight of the drain cover is 400 N and its centre of gravity is 0.60 m from the end P.

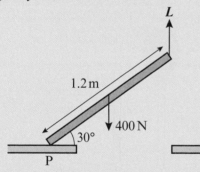

(a) Find the magnitude of the vertical lifting force, *L*, he applies to hold the cover steady in this position. **(3 marks)**

The cover is not moving (in equilibrium) so moments about P must sum to zero.

$L \times 1.2 \times \cos30° = 400 \times 0.60 \times \cos30°$

Therefore $L = 200\,N$

(b) Find the magnitude and direction of the contact force at the pivot P. **(3 marks)**

Consider vertical forces. $200 + P_v = 400$

Therefore the vertical component of *P* is 200 N

Consider horizontal forces. $P_h = 0$

Therefore the contact force at P is a vertical force of 200 N.

Now try this

(a) Look at the worked example. Calculate the vertical force the boy has to apply to start lifting the drain cover from its horizontal position. **(2 marks)**

(b) Calculate the force the boy would have to apply to the drain cover to hold it in the position shown (30° to the horizontal) if, instead of applying the force vertically, he applied it at 90° to the drain cover. **(2 marks)**

(c) Describe qualitatively how the contact force at P is affected as a result of the change in direction of the lifting force in (b). **(3 marks)**

Density and pressure

Physicists study matter under a wide range of densities and pressures, from the extreme conditions just after the Big Bang to the near-total vacuum of intergalactic space.

Density

$$\text{Density} = \frac{\text{mass}}{\text{volume}}$$

$$\rho = \frac{m}{V}$$

Rho (ρ) is the symbol for density. The S.I. unit is $kg\,m^{-3}$. A common alternative unit is $g\,cm^{-3}$.

$10\,kg\,m^{-3} = 0.001\,g\,cm^{-3}$

$1\,g\,cm^{-3} = 1000\,kg\,m^{-3}$

Pressure

Pressure = normal force/area

$$p = \frac{F}{A}$$

The S.I. unit is $N\,m^{-2}$ or Pa (pascal).

Pressure in a liquid

The pressure in a liquid or gas acts in all directions.

The pressure in a particular liquid depends only on depth and is calculated using

$$p = h\rho g$$

Densities of common materials

It is helpful to have a rough idea of the density of some common substances, especially when making estimates of physical quantities.

Substance	Density ($kg\,m^{-3}$)
air*	1.2
pure water	1000
seawater	1030 (at surface)
steel	8000 (varies)
mercury	13 600

*At room temperature and pressure.

Most materials expand as temperature rises, so density falls with increasing temperature. Water is an exception: in the range 0°C to 4°C, it contracts as it warms, so ice is actually less dense than very cold water and floats on top of it.

Measuring pressure differences

This equation can be used with a manometer to measure pressure differences. Here, a manometer is being used to measure the excess pressure inside a balloon compared with atmospheric pressure.

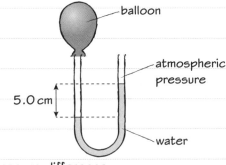

balloon
atmospheric pressure
5.0 cm
water

The pressure difference
$$\Delta p = \rho g \Delta h = 1000 \times 9.81 \times 0.05 = 490\,Pa$$

Note that this is the **excess** pressure, above atmospheric pressure, so the **total** pressure inside the balloon is about 100 490 Pa.

To measure pressure...

...use:

- pressure sensors and a datalogger
- a manometer (U-tube)
- a Bourdon gauge
- a barometer (for atmospheric pressure).

Worked example

A cubic block of steel has sides of length 5.0 cm.

(a) Calculate its mass and weight. (Use data from the table above and take $g = 9.81\,N\,kg^{-1}$.) **(3 marks)**

$V = 0.050^3 = 0.000125\,m^3$

$m = \rho V = 0.000125 \times 8000 = 1.0\,kg$

$W = mg = 1.0 \times 9.81 = 9.81\,N$

(b) The block is placed on a horizontal surface. Calculate the pressure due to the weight of the block on the surface. **(2 marks)**

The area of contact is $25\,cm^2 = 0.0025\,m^2$.

The pressure $p = \dfrac{F}{A} = \dfrac{9.81}{0.0025} = 3920\,Pa$.

Now try this

1 Divers claim that in the sea the excess pressure under water is approximately 1 atmosphere for every extra 10 m of depth. Show that this is approximately true. **(2 marks)**

2 Show that atmospheric pressure (about 10^5 Pa) is approximately equivalent to the pressure exerted at the base of a column of mercury of height 760 mm. **(2 marks)**

Upthrust and Archimedes' principle

The buoyant force that allows you to float in a swimming pool is caused by the variation of pressure with depth in a fluid.

Upthrust

Pressure in a fluid increases with depth, so when an object floats or is submerged in a fluid the pressure beneath the object is greater than the pressure above it. This creates an upward force on the object: the **upthrust**.

The volume of the floating or submerged object displaces an equal volume of the fluid.

Archimedes' principle states that upthrust is equal to the weight of the fluid displaced by the object.

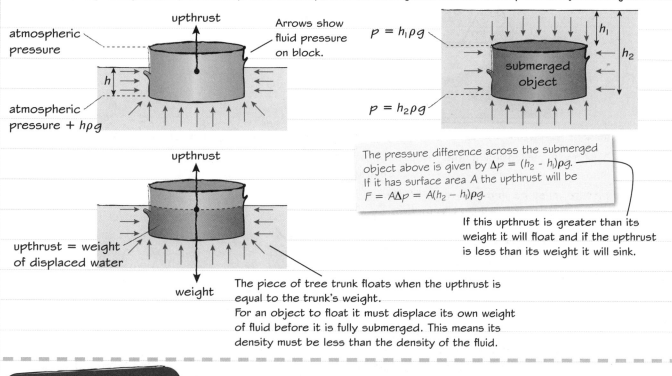

atmospheric pressure

upthrust

Arrows show fluid pressure on block.

$p = h_1 \rho g$

h

atmospheric pressure + $h\rho g$

h_1

submerged object

h_2

$p = h_2 \rho g$

upthrust

upthrust = weight of displaced water

weight

The pressure difference across the submerged object above is given by $\Delta p = (h_2 - h_1)\rho g$. If it has surface area A the upthrust will be $F = A\Delta p = A(h_2 - h_1)\rho g$.

If this upthrust is greater than its weight it will float and if the upthrust is less than its weight it will sink.

The piece of tree trunk floats when the upthrust is equal to the trunk's weight.
For an object to float it must displace its own weight of fluid before it is fully submerged. This means its density must be less than the density of the fluid.

Worked example

An oil tanker has a mass of 8.0×10^7 kg. Assume that its shape is approximately that of a rectangular box 250 m long and 75 m wide. The density of seawater is 1030 kg m^{-3}.

(a) What volume of water must it displace in order to float? **(3 marks)**

The weight of seawater displaced must equal the weight of the boat. This means that the mass of water displaced is equal to the mass of the boat.

Therefore $V = \dfrac{m}{\rho} = \dfrac{8.0 \times 10^7}{1030} = 7.8 \times 10^4$ m^3.

(b) How far below the surface of the sea is the bottom of the ship when it floats? **(3 marks)**

The ship is a rectangular block, so the displaced water would also be a rectangular block with base area 250 × 75 = 18750 m^2. The height of this block (the depth of the bottom of the boat) is therefore

$h = \dfrac{\text{volume}}{\text{area}} = \dfrac{7.8 \times 10^4}{18750} = 4.16$ m.

Now try this

1 Look at the expression for the upthrust on the submerged object above: $F = A(h_2 - h_1)\rho_{\text{fluid}}g$. Explain why it will sink if its density ρ_{object} is greater than that of the fluid, ρ_{fluid}, and float if ρ_{object} is less than ρ_{fluid}. **(4 marks)**

2 Look at the worked example about the oil tanker.

(a) Calculate the pressure due to seawater on the bottom of the boat. **(2 marks)**

(b) Show that the pressure due to the seawater provides an upthrust equal to the weight of the boat. **(3 marks)**

Exam skills

This exam-style question uses knowledge and skills you have already revised. Have a look at pages 17–19 for a reminder about forces, moments and equilibrium.

Worked example

The diagram below shows a picture hanging by two strings from a hook on a wall. The tension in each string is **T**.

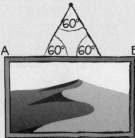

The mass of the picture is 920 g.

(a) Explain why the tension in both strings must be equal. **(2 marks)**

The picture must be in equilibrium, so the horizontal forces exerted on the picture by each string must be equal and opposite. The horizontal component of tension from one string is $T\cos 60°$. They are both at the same angle so must have the same tension.

(b) Calculate the tension in one of the strings. **(3 marks)**

The picture is in equilibrium so the total upward force (2T, twice the tension in each string) must balance the weight of the picture.

$2T\sin 60° = mg$

$T = \dfrac{mg}{2\sin 60°} = 0.920 \times \dfrac{9.81}{2\sin 60°} = 5.21\,N$

(c) State and explain what would happen to the tension in the strings if they were connected to the corners A and B of the picture rather than the positions shown. **(3 marks)**

The tension would increase. $T = \dfrac{mg}{2\sin\theta}$, where θ is the angle to the horizontal. When the strings are moved to the corners of the picture, θ gets smaller so **T** must increase.

(d) Explain why the centre of mass of the picture must lie vertically below the supporting nail. **(2 marks)**

When the centre of mass is directly below the nail the line of action of the weight acts through the nail. The lines of action of the forces in both strings also act through the nail. If all three forces acting on the picture pass through the same point then there is no resultant moment (no turning effect) so the moments are in equilibrium.

Command words: State and explain

Read the question carefully – when you are asked to state and explain, there will usually be marks for both parts of your answer.

Remember that for an object to be in equilibrium both horizontal **and** vertical forces must add to zero.

Maths skills Make sure you show each part of your working when carrying out a calculation. Usually this means you must:

✓ state the equation

✓ rearrange the equation

✓ substitute values from the question

✓ calculate an answer.

Take care to convert non-S.I. units to S.I. units before carrying out your calculations. In this case the mass is given in grams; you must convert to kilograms in order to get a weight in newtons.

Maths skills If you introduce a new variable, such as θ, it is important to state what it represents. Here it is the angle of the string to the horizontal.

You might be tempted to answer this part by simply saying that this is necessary for the picture to be in equilibrium. A better answer would be to explain that there is no resultant moment acting on the picture so the moments are in equilibrium.

There are often alternative ways to answer a question. In this case instead of describing the equilibrium condition you could explain that if the centre of mass was not vertically beneath the nail then there would be resultant moment acting on the picture so it would not be in equilibrium.

Work done by a force

When a force moves something it does work on it. This simple idea is the foundation for understanding energy transfers.

Work

Work done is energy transferred.

Work done = force × distance moved in the direction of the force

work done = Fx (displacement parallel to force)

work done = $Fx\cos\theta$ (displacement at angle θ to force)

When A exerts a force on B and moves it, the work done by the force transfers an equal amount of energy from A to B.

Forces that do not cause movement do no work. For example, when a book rests on a table, gravity continues to pull it downward but does no work on it. If the book falls off the table then the gravitational force moves, and does work on, the book, transferring gravitational potential energy (E_P) into kinetic energy (E_K) as it falls.

Units for work and energy

Work and energy are measured in joules (J), the same as N m.

Definition of the **joule**:

1 J is the work done by a force of 1 N when it moves an object 1 m parallel to its line of action.

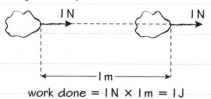

work done = 1 N × 1 m = 1 J

Gravitational potential energy

To lift something vertically through a height h you must exert an upward force equal to the object's weight, mg.

Work done = $mg \times h$ Learn this derivation.

This energy is transferred from you and stored as gravitational potential energy in the body.

$\Delta E_P = mgh$ (assuming g remains constant)

Kinetic energy

To accelerate an object from rest to velocity v, you need to apply a force to the object. If this is done with a constant force, F, and the object moves a distance s as it accelerates then:

$v^2 = u^2 + 2as \quad a = \dfrac{F}{m} \quad u = 0$

$v^2 = \dfrac{2Fs}{m}$ Learn this derivation.

The work done is $Fs = \frac{1}{2}mv^2$

This work has transferred energy from you to become kinetic energy in the moving body.

$E_K = \frac{1}{2}mv^2$

Worked example

(a) The thrust from a model rocket engine is 25 N. The rocket rises through a distance of 30 m before the motor cuts out.

How much work is done on the rocket? **(1 mark)**

$W = Fx = 25 \times 30 = 750\,\text{J}$

(b) A boat is pulled forwards by two tugs. The cables connecting the tugs to the boat each make an angle of 20° to the direction of motion of the boat. The tension in each cable is 20 000 N. How much work is done on the boat when it moves forwards by 25 m? **(2 marks)**

Work done by one of the forces
= 20 000 × 25 × cos20° = 470 000 J

There are two cables, therefore the total work done on the boat is 940 000 J.

Now try this

1 The average braking force on a car is 5000 N, and when the brakes are applied it stops in a distance of 80 m.
   (a) How much work is done? **(2 marks)**
   (b) Describe the energy transfer that takes place. **(1 mark)**

2 A child uses a string to pull a sledge 20 m across snow. The average force applied by the child is 30 N and the string makes an angle of 35° to the horizontal. Calculate the work done by the child. **(2 marks)**

Conservation of energy

The Universe is constantly changing, but the total energy in it always remains the same.

The law of conservation of energy

This law can be stated in different ways.

- Energy can neither be created nor destroyed; it can only be transferred from one form to another.
- The total energy of the Universe is constant.
- The total energy of a closed system remains constant.

A **closed system** is one which does not exchange energy with its surroundings.

Energy transfers

- Work – energy transferred by the action of a force.
- Heating – energy transferred because of a temperature difference.

Energy flow diagrams (Sankey diagrams)

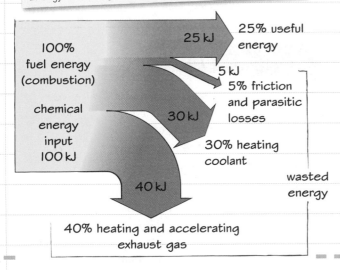

Energy flow diagram for an internal combustion engine

100% fuel energy (combustion)

chemical energy input 100 kJ

25 kJ → 25% useful energy

5 kJ 5% friction and parasitic losses

30 kJ → 30% heating coolant

40 kJ → 40% heating and accelerating exhaust gas

wasted energy

These diagrams can also be drawn in terms of power instead of energy. If the engine above transfers 100 kJ in 1 s then all the energy values in kJ could be replaced by power values in kW.

Calculating efficiency from the energy flow diagram

$$\text{Efficiency} = \frac{\text{(useful energy output)}}{\text{(total energy input)}} \times 100\,\%$$

or

$$\text{Efficiency} = \frac{\text{(useful power output)}}{\text{(total power input)}} \times 100\,\%$$

Forms of energy

- Gravitational potential energy – the energy a body has because of its height in a gravitational field.
- Kinetic energy – the energy a body has because of its motion.
- Elastic potential energy (strain energy) – the energy stored in a body when it is stretched or compressed.
- Thermal energy (heat) – the energy a body has because of the random thermal motions of its particles.
- Electromagnetic energy – energy transferred by electromagnetic waves such as light.
- Sound energy – energy transferred by sound waves.
- Electrical energy – energy transferred as a result of charges moving through a potential difference.
- Chemical energy – the energy released or absorbed when chemical bonds form or break.
- Nuclear energy – the energy released when nucleons bind together.

Worked example

Look at the energy flow diagram.

(a) How does it show that energy is conserved?

(2 marks)

The total energy input is equal to the total energy output:
100 kJ in = 25 kJ + 5 kJ + 30 kJ + 40 kJ out.

(b) What percentage of input energy is wasted?

(2 marks)

5 kJ + 30 kJ + 40 kJ = 75 kJ lost

$\frac{75}{100} \times 100\% = 75\%$ lost.

Now try this

Look at the energy flow diagram for the internal combustion engine.

(a) What are the main energy transfers taking place?

(3 marks)

(b) What is the efficiency of the engine?   **(1 mark)**

(c) When this engine is used to drive a car at constant speed along a level motorway the kinetic energy of the car is constant but the engine continues to produce 25 kW of useful output power – explain how energy is conserved in this case.   **(3 marks)**

Kinetic and gravitational potential energy

In many situations kinetic energy is transferred to gravitational potential energy and vice versa. The principle of conservation of energy gives us a useful way of solving problems about these situations.

Equations

Kinetic energy $E_K = \frac{1}{2}mv^2$

Change in gravitational potential energy
$\Delta E_P = mg\Delta h$ — This equation can **only** be used when an object moves in a region where g is constant.

Conservation of energy

If there are no frictional forces then the total energy $E_K + E_P$ is constant. If kinetic energy is lost, there will be an equal gain in gravitational potential energy, and vice versa. This is useful in many calculations.

Total $E = E_K + E_P$

Energy conservation in space

Satellites, moons and planets move through space where there is no friction so their total energy $E_P + E_K$ is constant. Look at the diagram below, which shows the Moon's elliptical orbit around the Earth.

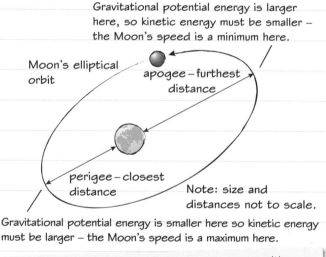

Gravitational potential energy is larger here, so kinetic energy must be smaller – the Moon's speed is a minimum here.

Moon's elliptical orbit

apogee – furthest distance

perigee – closest distance

Note: size and distances not to scale.

Gravitational potential energy is smaller here so kinetic energy must be larger – the Moon's speed is a maximum here.

In this example g is not constant, as it decreases with distance from the Earth's surface. Nevertheless gravitational potential energy still increases as an object moves away from the body attracting it.

E_P TO E_K

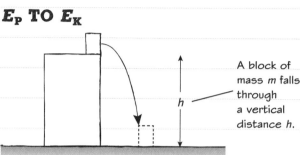

A block of mass m falls through a vertical distance h.

Initial $E_P = mgh$ Initial $E_K = 0$

Final $E_P = 0$ Final $E_K = \frac{1}{2}mv^2$

If there are no losses due to friction then the total energy $E_P + E_K$ is constant

Gain of E_K = loss of E_P

$$\frac{1}{2}mv^2 = mgh$$
$$v = \sqrt{2gh}$$

Worked example

The diagram below shows part of a roller coaster. Each car with its passengers has a mass of 500 kg and is winched up to point A before being released to run freely along the rest of the track.

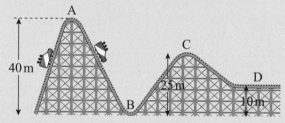

(a) Calculate the change in gravitational potential energy for each car as it is winched up to point A. **(1 mark)**

$\Delta E_P = mg\Delta h = 500 \times 9.81 \times 40 = 196\,200\,\text{J}$

(b) Calculate the maximum velocity of a car as it reaches B and explain why the actual velocity at B will be lower than this. **(3 marks)**

If there are no frictional losses then the gain of kinetic energy from A to B will equal the loss of gravitational potential energy from A to B.

$$\frac{1}{2}mv^2 = mgh$$
$$v = \sqrt{2gh} = \sqrt{2 \times 9.81 \times 40} = 28\,\text{m s}^{-1}$$

In reality there will be frictional forces acting against the motion of the car. Some of the gravitational potential energy will be transferred by heating so the kinetic energy and therefore the speed at B will be lower.

Now try this

Look at the diagram of the roller coaster.

(a) What is the minimum speed that the car could have at B and still reach C? **(2 marks)**

(b) Why, in practice, must the speed at B be greater than the value you have calculated in A if the car is to reach C? **(2 marks)**

(c) If the car pauses briefly at C before rolling downward again, what is the maximum speed that it could have at D? **(2 marks)**

Mechanical power and efficiency

In physics, power means the rate of energy transfer – the faster energy is transferred, the more power.

Power

Power is the rate of transfer of energy.

The S.I. unit of power is the watt (W).

A power of 1 watt is equivalent to a rate of energy transfer of 1 joule per second.

$$\text{Power (W)} = \frac{\text{energy transfer (J)}}{\text{time (s)}} \qquad P = \frac{E}{t}$$

or

$$\text{Power (W)} = \frac{\text{work done (J)}}{\text{time taken (s)}} \qquad P = \frac{W}{t}$$

Common units of power

- 1 milliwatt = 1 mW = 0.001 W = 10^{-3} W
- 1 kilowatt = 1 kW = 1000 W = 10^{3} W
- 1 megawatt = 1 MW = 1 000 000 W = 10^{6} W
- 1 gigawatt = 1 GW = 1 000 000 000 W = 10^{9} W

Power of a moving force

The diagram shows a force acting on an object and moving it forwards. The dotted box shows its new position a short time, δt, later. During this short time it has moved a distance δs.

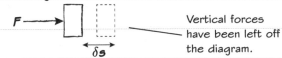

The work done in this time is $\delta W = F\delta s$

The power $P = \dfrac{\delta W}{\delta t} = \dfrac{F\delta s}{\delta t} = Fv$

where v is the velocity ($= \dfrac{\delta s}{\delta t}$).

Power (W) = force (N) × velocity (m s^{-1})

$$P = Fv \qquad \text{Learn this derivation.}$$

Power and drag

When a car is travelling at constant velocity, v, the forces acting on it are balanced (in equilibrium). This means the driving force, F, is equal to the total drag force, D, and the resultant is zero.

Vertical forces have been left off the diagram.

The power supplied to the car is $P = Fv$

but $F = D$, so $P = Dv$

Drag force $D = \dfrac{P}{v}$

Efficiency

Efficiency is a measure of how effectively a transducer transfers energy into another useful form. The higher the efficiency, the higher the proportion of useful output energy.

$$\text{Efficiency} = \frac{\text{(useful energy output)}}{\text{(total energy input)}} \times 100\,\%$$

or

$$\text{Efficiency} = \frac{\text{(useful power output)}}{\text{(total power input)}} \times 100\,\%$$

A 70 kg man runs up a flight of stairs, rising 4.0 m vertically in a time of 5.0 s.

(a) Calculate his average useful power output in climbing the stairs. **(3 marks)**

Work done = mgh = 70 × 9.81 × 4.0
= 2746.8 J

Power = $\dfrac{\text{work done}}{\text{time taken}} = \dfrac{2746.8}{5} = 550$ W

(b) If his muscles are only 20% efficient, at what rate did he transfer chemical energy as he climbed the stairs? **(2 marks)**

The rate of transferring chemical energy is the power input.

$\dfrac{550}{P_{chem}} = 0.20$

$P_{chem} = \dfrac{550}{0.20} = 2750$ W

(c) A car is driving at a constant speed of 20 m s^{-1} along a straight road. The useful output power of the engine is 250 kW. Calculate the drag force acting on the car. **(3 marks)**

The useful output power is used to do work against the frictional forces and heats the surroundings.

$P = Fv$ so $F = \dfrac{P}{v} = \dfrac{250\,000}{20} = 12\,500$ N

Since the car is moving forward at constant speed the total horizontal force acting on it is zero, so the forward force is equal to the drag force D.

$D = 12\,500$ N.

A boat is moving at a constant velocity of 2.0 m s^{-1}. The useful power output of the boat's diesel engine is 50 kW. Calculate the drag force on the boat. **(3 marks)**

Exam skills

This exam-style question uses knowledge and skills you have already revised. Have a look at pages 23–26 for a reminder about forces, energy and motion.

Worked example

A student wants to investigate the transfer of gravitational potential energy to kinetic energy as a toy car moves down a slope. She does this by releasing the car from rest and measuring the speed, v, when it has moved a distance, d, along the slope. She records the speed of the car for a range of values of d and then plots a graph of v^2 against d.

(a) Describe an experimental method to measure the speed of the car when it has moved a distance, d.
(2 marks)

> Note the command word 'describe'. This means that you should not just write 'use a light gate' – explain how it is used.

Attach a card of measured length (e.g. 5.0 cm) to the top of the car and place a light gate at distance d. When the card cuts the light beam a data logger can use the time and length of card to calculate and display the speed.

> Questions like this do not have one set answer. You could use a motion sensor or video capture and analysis software instead of a light gate, but whatever you choose to use you must describe **how** the apparatus is set up and used, and the data that it measures.

(b) Show that a graph of v^2 against d should be a straight line through the origin, if work done against friction can be neglected. **(4 marks)**

Loss of GPE = gain of KE

$mgh = \frac{1}{2}mv^2$

h = the height of the ramp, g the acceleration due to gravity, m the mass of the car and v its measured velocity.

$mgd\sin\theta = \frac{1}{2}mv^2$

$v^2 = (2g\sin\theta)d$

All terms inside the bracket are constant so v^2 is directly proportional to d, so the graph will be a straight line through the origin.

> **Maths skills** To answer part (b) you need to:
> 1. start with conservation of energy
> 2. write down the equation and define the terms
> 3. rearrange the equation
> 4. show that it leads to $v^2 \propto d$.

(c) Use these data to calculate the average frictional force, F, acting on the car:
mass of car = 0.40 kg
angle of slope to horizontal = 20°
distance d = 45 cm
speed v = 1.65 m s⁻¹ **(3 marks)**

> 1. Start by stating the underlying physics – in this case it is the law of conservation of energy.
> 2. Use it to work out the total work done against friction.
> 3. Finally use this intermediate result and the equation $W = Fd$ to calculate a value for F.

The total input energy is from GPE. Since energy is conserved, any energy that is not transferred to KE must be transferred to heat by work done against friction.

Work done against friction

$= $ loss of GPE $-$ gain in KE $= mgd\sin\theta - \frac{1}{2}mv^2$

$= 0.4 \times 9.81 \times 0.45 \times \sin 20° - 0.4 \times \dfrac{(1.65)^2}{2}$

$= 0.604 - 0.545 = 0.059 \text{ J}$

Work done against friction $= Fd$

$F = \dfrac{0.059}{0.45} = 0.13 \text{ N}$

Elastic and plastic deformation

Bend a thick metal wire slightly and it will ping back into shape – bend it a lot and it will stay bent.

Compression and tension

Here are some terms used when forces **deform** samples of material by squashing (compression) or pulling (tension):

extension (positive extension) = stretched length − original length

compression (negative extension) = original length − compressed length

tensile deformation: forces acting along the axis of the sample tend to extend it

compressive deformation: forces acting along the axis of the sample tend to compress it

elastic deformation: the object returns to its original shape and size when the applied forces are removed. There is no permanent deformation.

plastic deformation: the object is permanently deformed even when the applied forces are removed.

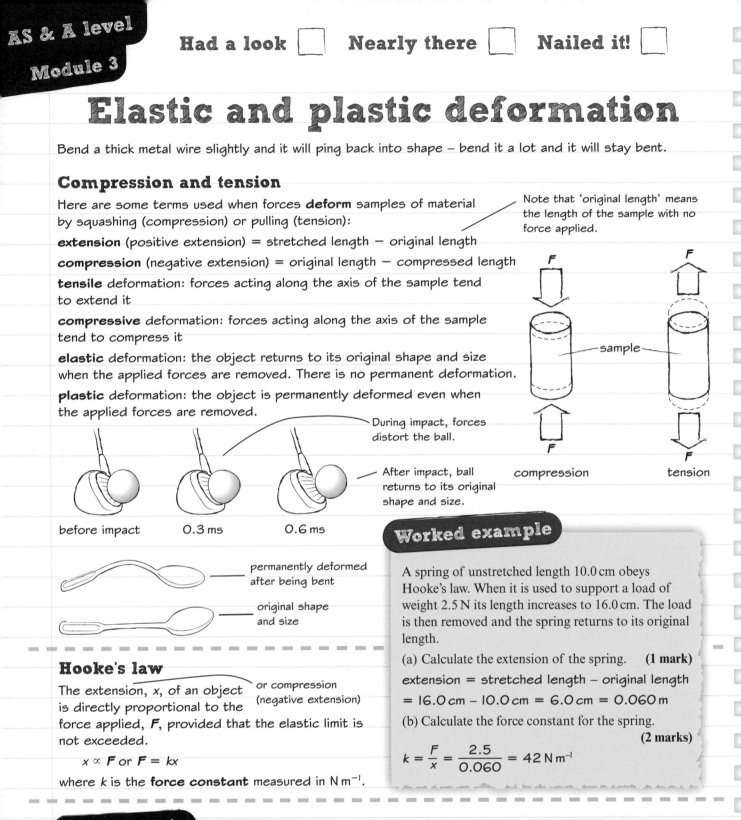

Note that 'original length' means the length of the sample with no force applied.

During impact, forces distort the ball.

After impact, ball returns to its original shape and size.

compression   tension

before impact   0.3 ms   0.6 ms

permanently deformed after being bent

original shape and size

Hooke's law

The extension, x, of an object is directly proportional to the force applied, F, provided that the elastic limit is not exceeded.

or compression (negative extension)

$$x \propto F \text{ or } F = kx$$

where k is the **force constant** measured in $N\,m^{-1}$.

A spring of unstretched length 10.0 cm obeys Hooke's law. When it is used to support a load of weight 2.5 N its length increases to 16.0 cm. The load is then removed and the spring returns to its original length.

(a) Calculate the extension of the spring.   **(1 mark)**

extension = stretched length − original length

= 16.0 cm − 10.0 cm = 6.0 cm = 0.060 m

(b) Calculate the force constant for the spring.

(2 marks)

$$k = \frac{F}{x} = \frac{2.5}{0.060} = 42\,N\,m^{-1}$$

Now try this

A load of weight 4.00 N is suspended from a steel spring and then removed. The process is then repeated with a sequence of loads of increasing weight. The length of the spring when each load is suspended and removed is recorded in the table below.

Load (N)	0.00	4.00	8.00	12.00
Spring length – with load (cm)	14.0	17.0	20.0	25.0
Spring length – load removed (cm)	14.0	14.0	14.0	17.0

(a) How can you tell that the spring behaves elastically when a load of weight 8.00 N is suspended from it but plastically when a load of weight 12.00 N is suspended from it?   **(2 marks)**

(b) It is suggested that this spring obeys Hooke's law up to a load of weight 8.00 N but not beyond that. Explain how the data support these claims.   **(2 marks)**

(c) Use the data for a load of 8.00 N to calculate the force constant of the spring.   **(3 marks)**

Stretching things

Practical skills Engineers need to have ways of predicting how materials will behave under all kinds of applied forces, like tension.

Investigating the extension of a spring

The red mark on the ruler is a **fiducial mark**, a reference point for the observer.

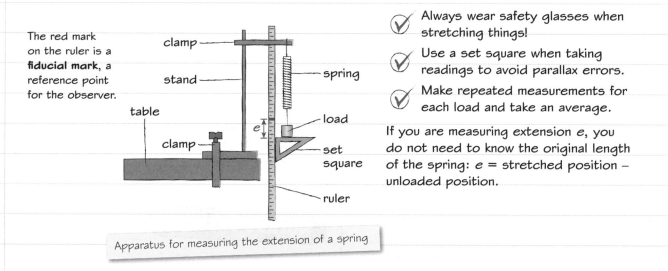

✓ Always wear safety glasses when stretching things!

✓ Use a set square when taking readings to avoid parallax errors.

✓ Make repeated measurements for each load and take an average.

If you are measuring extension e, you do not need to know the original length of the spring: e = stretched position − unloaded position.

Apparatus for measuring the extension of a spring

Investigating the extension of a thin metal wire

The extension is $e = l - l_0$

l = stretched length

l_0 = unstretched (original) length

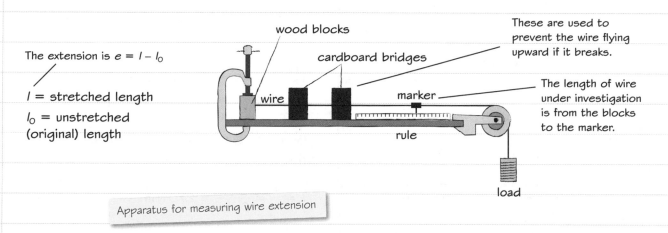

These are used to prevent the wire flying upward if it breaks.

The length of wire under investigation is from the blocks to the marker.

Apparatus for measuring wire extension

Now try this

1 Look at the experiment on stretching a spring.

(a) Explain how a parallax error might occur if measurements were made without the set square. **(2 marks)**

(b) What type of error is a parallax error? **(1 mark)**

2 Look at the experiment to stretch a thin metal wire.

(a) Explain why it is a good idea to use a long wire. **(2 marks)**

(b) What instrument would be suitable to measure the length of the stretched wire? **(1 mark)**

(c) What instrument would be suitable to measure the diameter of the wire? **(1 mark)**

(d) What precautions should be taken to ensure accuracy when making the measurements in (b) and (c)?

(3 marks)

Force-extension graphs

Graphs of force against extension tell us about the behaviour of springs and wires.

Interpreting a force-extension graph

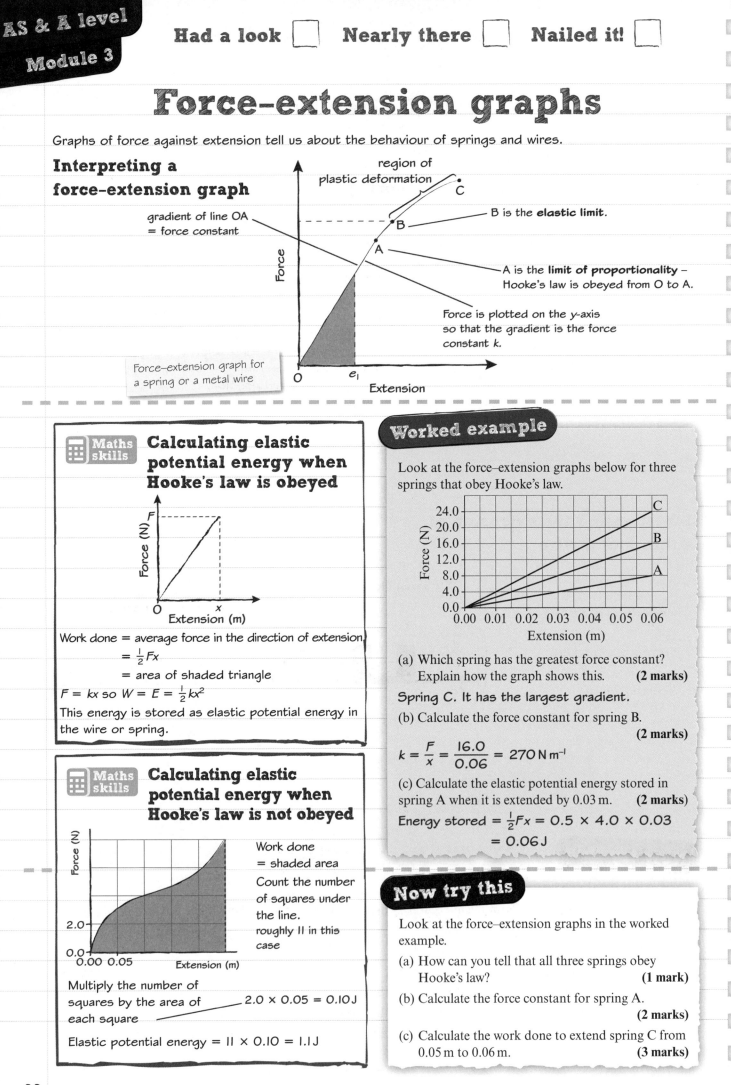

region of plastic deformation

gradient of line OA = force constant

B is the **elastic limit**.

A is the **limit of proportionality** – Hooke's law is obeyed from O to A.

Force is plotted on the y-axis so that the gradient is the force constant k.

Force–extension graph for a spring or a metal wire

Force

O e_1 Extension

🖩 Maths skills Calculating elastic potential energy when Hooke's law is obeyed

Force (N)
F

O   Extension (m)   x

Work done = average force in the direction of extension.
= $\frac{1}{2}Fx$
= area of shaded triangle
$F = kx$ so $W = E = \frac{1}{2}kx^2$
This energy is stored as elastic potential energy in the wire or spring.

🖩 Maths skills Calculating elastic potential energy when Hooke's law is not obeyed

Force (N)

2.0

0.0
0.00 0.05   Extension (m)

Work done
= shaded area
Count the number of squares under the line.
roughly 11 in this case

Multiply the number of squares by the area of each square   2.0 × 0.05 = 0.10 J

Elastic potential energy = 11 × 0.10 = 1.1 J

Worked example

Look at the force–extension graphs below for three springs that obey Hooke's law.

Force (N)
24.0
20.0
16.0
12.0
8.0
4.0
0.0
0.00 0.01 0.02 0.03 0.04 0.05 0.06
Extension (m)

C
B
A

(a) Which spring has the greatest force constant? Explain how the graph shows this.   **(2 marks)**

Spring C. It has the largest gradient.

(b) Calculate the force constant for spring B.
   (2 marks)

$k = \dfrac{F}{x} = \dfrac{16.0}{0.06} = 270\,\text{N m}^{-1}$

(c) Calculate the elastic potential energy stored in spring A when it is extended by 0.03 m.   **(2 marks)**

Energy stored = $\frac{1}{2}Fx$ = 0.5 × 4.0 × 0.03
= 0.06 J

Now try this

Look at the force–extension graphs in the worked example.

(a) How can you tell that all three springs obey Hooke's law?   **(1 mark)**

(b) Calculate the force constant for spring A.
   (2 marks)

(c) Calculate the work done to extend spring C from 0.05 m to 0.06 m.   **(3 marks)**

Stress and strain

If we want to discuss the properties of materials (e.g. steel) rather than things (e.g. a steel spring) we must use the intrinsic quantities of stress and strain rather than force and extension.

Stress

Tensile stress = $\dfrac{\text{force}}{\text{cross-sectional area}}$

$$\sigma = \frac{F}{A}$$

S.I. units are $N\,m^{-2}$ or Pa (pascal).

Stresses are often large numerical values, so we also use MPa ($= 10^6$ Pa) and GPa ($= 10^9$ Pa).

Strain

Tensile strain = $\dfrac{\text{extension}}{\text{original length}}$

$$\varepsilon = \frac{e}{l}$$

This is a ratio of lengths, so it has no unit (it is dimensionless).

Stress versus strain graph for a metal wire

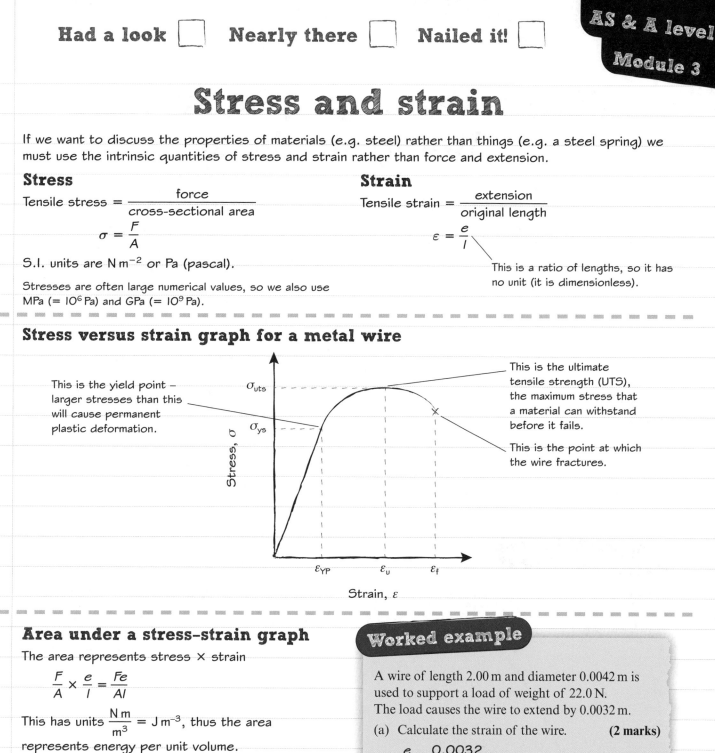

This is the yield point – larger stresses than this will cause permanent plastic deformation.

This is the ultimate tensile strength (UTS), the maximum stress that a material can withstand before it fails.

This is the point at which the wire fractures.

Area under a stress–strain graph

The area represents stress × strain

$$\frac{F}{A} \times \frac{e}{l} = \frac{Fe}{Al}$$

This has units $\dfrac{N\,m}{m^3} = J\,m^{-3}$, thus the area represents energy per unit volume.

Properties of materials

The beauty of using stress and strain is that the values are for the material and not for a particular object. For example, the ultimate tensile strength of copper is 220 MPa whether the copper is formed into a fine wire or a thick bar; it makes no difference.

Worked example

A wire of length 2.00 m and diameter 0.0042 m is used to support a load of weight of 22.0 N. The load causes the wire to extend by 0.0032 m.

(a) Calculate the strain of the wire. **(2 marks)**

$$\varepsilon = \frac{e}{l} = \frac{0.0032}{2.00} = 0.0016$$

(b) Calculate the stress in the wire. **(4 marks)**

Cross-sectional area of wire

$$= \pi\left(\frac{d}{2}\right)^2 = \pi \times (0.0021)^2 = 1.385 \times 10^{-5}\,m^2$$

$$\text{Stress} = \frac{F}{A} = \frac{22.0}{1.385 \times 10^{-5}} = 1.59 \times 10^6\,Pa$$

$$= 1.59\,MPa$$

Now try this

1 A wire of diameter 0.0033 m and length 1.80 m extends to a length 1.81 m when pulled by a tensile force of 6.2 N.

  (a) Calculate the strain in the wire. **(2 marks)**

  (b) Calculate the stress in the wire. **(4 marks)**

2 Copper has an ultimate tensile strength of 220 MPa. What is the maximum force that a copper wire of diameter 0.20 mm can withstand? **(4 marks)**

Stress–strain graphs and the Young modulus

A stress–strain graph illustrates the mechanical properties of a material.

Hooke's law and the Young modulus

Hooke's law is obeyed when the stress–strain graph is a straight line that passes through the origin.

In this region stress is directly proportional to strain: $\sigma \propto \varepsilon$

The constant of proportionality is E, the **Young modulus**.

$$\sigma = E\varepsilon$$

The Young modulus is a measure of the stiffness of the material. It is equal to the gradient of the stress–strain graph and has units $N\,m^{-2}$ or Pa.

In the example above $E = \dfrac{240 \times 10^6}{0.004} = 60\,GPa$

Materials terminology

Strong: large value of ultimate tensile strength.

Stiff: high Young modulus – low strain for high stress (a steep initial gradient in the stress–strain graph).

Plastic: applied stress causes permanent deformation through permanent internal structural change.

Ductile: can be drawn out into wires – ductile materials must be plastic.

Tough: absorbs a lot of energy before fracture – tough materials have a large area under the stress–strain graph.

Brittle: undergoes little or no plastic deformation before fracture and then breaks suddenly as cracks move rapidly through the material.

Elastic: deformation is reversible – material returns to original shape and size when stress is removed.

Worked example

Look at the stress–strain graph below.

$E = 95.2\,GPa$

(a) Over what range, approximately, does this material obey Hooke's law? **(2 marks)**

This is the region in which the graph is a straight line through the origin – from a stress of 0 Pa to about 300 MPa.

(b) Calculate the Young modulus of this material. **(4 marks)**

This is the gradient of the line in the straight section. Reading from the graph, $\sigma = 300\,MPa$ when $\varepsilon = 0.0033$.

Therefore $E = \dfrac{300 \times 10^6}{0.0033} = 9 \times 10^{10}\,Pa$

$= 90\,GPa$ (1 significant figure)

Stress–strain graphs for materials with different mechanical properties

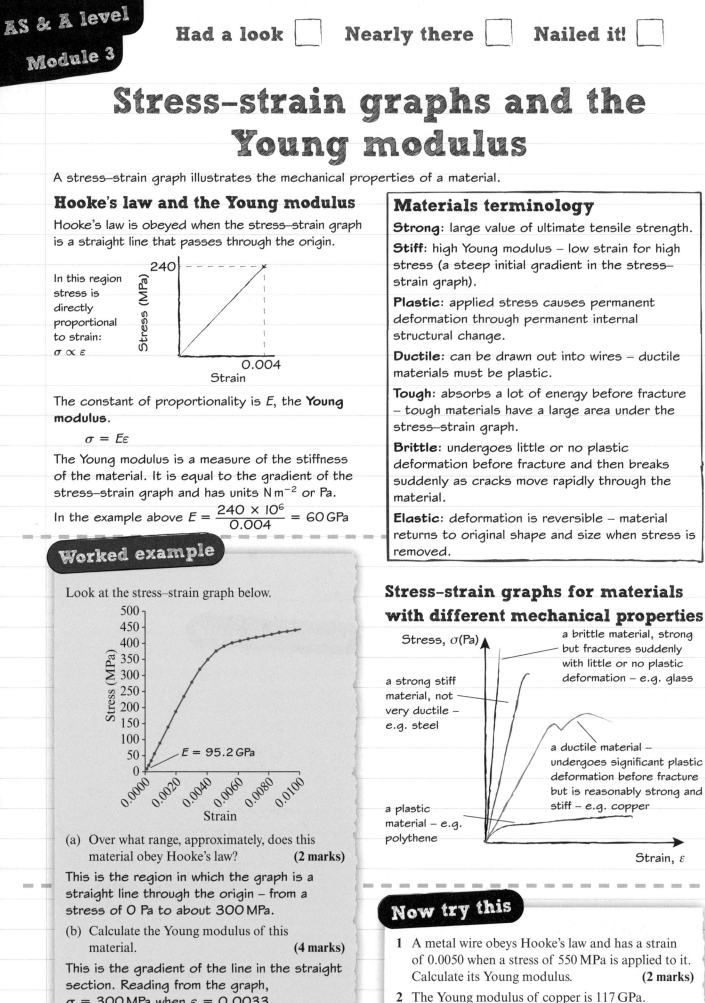

a brittle material, strong but fractures suddenly with little or no plastic deformation – e.g. glass

a strong stiff material, not very ductile – e.g. steel

a ductile material – undergoes significant plastic deformation before fracture but is reasonably strong and stiff – e.g. copper

a plastic material – e.g. polythene

Now try this

1 A metal wire obeys Hooke's law and has a strain of 0.0050 when a stress of 550 MPa is applied to it. Calculate its Young modulus. **(2 marks)**

2 The Young modulus of copper is 117 GPa. A copper wire of unstretched length 4.00 m and diameter 0.50 mm is used to suspend a mass of 10 kg. Calculate its extension. **(5 marks)**

Measuring the Young modulus

The Young modulus, E, measures the 'stiffness' of a material.

Practical skills — Measuring the Young modulus of a metal wire

$$E = \frac{\sigma}{\varepsilon} = \frac{F/A}{e/l}$$

- F is the force applied to the wire – hang known masses and use $F = mg$.
- A is the cross-sectional area of the wire – use a micrometer screw gauge to measure the wire diameter and then calculate area using $A = \pi\left(\dfrac{d}{2}\right)^2$.

- e is the extension of the wire – mark the original length of the wire and find e from l', l where l' is the extended length.
- l is the original length of the wire – use a metre ruler to measure l.

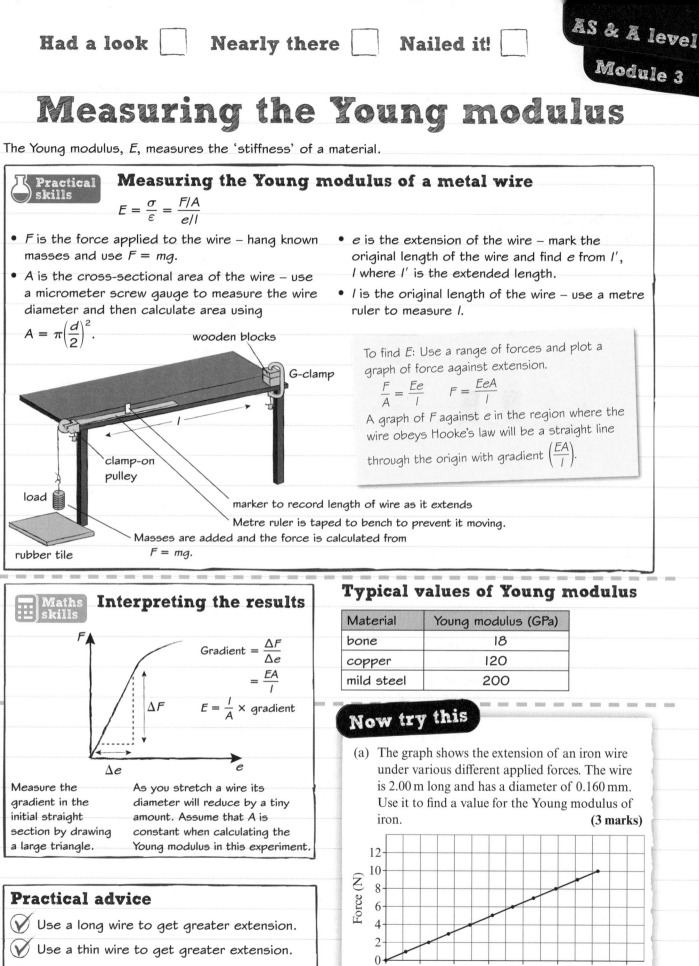

wooden blocks

G-clamp

clamp-on pulley

load

rubber tile

marker to record length of wire as it extends

Metre ruler is taped to bench to prevent it moving.

Masses are added and the force is calculated from $F = mg$.

To find E: Use a range of forces and plot a graph of force against extension.

$$\frac{F}{A} = \frac{Ee}{l} \qquad F = \frac{EeA}{l}$$

A graph of F against e in the region where the wire obeys Hooke's law will be a straight line through the origin with gradient $\left(\dfrac{EA}{l}\right)$.

Maths skills — Interpreting the results

$$\text{Gradient} = \frac{\Delta F}{\Delta e}$$
$$= \frac{EA}{l}$$
$$E = \frac{l}{A} \times \text{gradient}$$

Measure the gradient in the initial straight section by drawing a large triangle.

As you stretch a wire its diameter will reduce by a tiny amount. Assume that A is constant when calculating the Young modulus in this experiment.

Practical advice

- ✓ Use a long wire to get greater extension.
- ✓ Use a thin wire to get greater extension.
- ✗ Make sure the wire has no kinks at the start.
- ✓ Correct the micrometer measurements for zero errors.
- ✓ Wear safety glasses in case the wire breaks.

Typical values of Young modulus

Material	Young modulus (GPa)
bone	18
copper	120
mild steel	200

Now try this

(a) The graph shows the extension of an iron wire under various different applied forces. The wire is 2.00 m long and has a diameter of 0.160 mm. Use it to find a value for the Young modulus of iron. **(3 marks)**

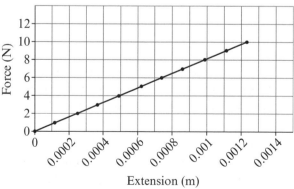

(b) State the actions you would take if carrying out this experiment to ensure an accurate result.

(2 marks)

33

Exam skills

This exam-style question uses knowledge and skills you have already revised. Have a look at pages 31–33 for a reminder about stress, strain and the Young modulus, and look at pages 3 and 4 for a reminder about how to work with uncertainties.

Worked example

A student carries out an experiment to measure the Young modulus of iron. He sets up his apparatus as shown below, using an iron wire. He then measures the length of the wire for a range of different loads. He also measures the diameter of the wire.

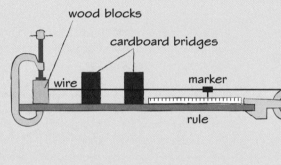

wood blocks
cardboard bridges
wire
marker
rule
load

Practical skills Make sure you know how to use a micrometer screw gauge to measure the diameter of a wire.
Repeat the measurement at least three times, at different points along the wire and at different orientations, and use the average value. This will reduce the effect of random measurement errors.

(a) Suggest a suitable piece of equipment for measuring the diameter d of the wire. Explain why this equipment is suitable. **(3 marks)**

A micrometer screw gauge is suitable because it can measure the diameter to the nearest 0.01 mm. This level of precision reduces the uncertainty in this measurement.

Remember that the precision of the measuring instrument is equal to the smallest scale division on the instrument (in this case 0.01 mm).

(b) Explain why using a longer wire could reduce the uncertainty in the final value of the Young modulus. **(2 marks)**

A longer wire will extend more under the same load. The measurement error will not change but it will be a smaller fraction of the actual measurement so the percentage uncertainty in the result will be reduced.

Look at pages 3 and 4 to review how measurement errors affect the uncertainty in your results.

(c) The student decides to plot a graph of stress against strain.

   (i) Explain how he can calculate the stress σ in the wire for a load of mass m. **(3 marks)**

Stress = force ÷ area, so he needs to divide the force provided by the load ($F = mg$) by the cross-sectional area of the wire ($A = \frac{1}{4}\pi d^2$).

$$\sigma = \frac{4mg}{\pi d^2}.$$

Maths skills Be careful here – it is very easy to use the diameter instead of the cross-sectional area or to mix up diameter and radius!

   (ii) Explain how he can calculate the strain in the wire for any particular load. **(2 marks)**

Strain is defined as extension ÷ original length. To find the extension he must subtract the original length from the extended length at this load, $e = l - l_0$.

Practical skills When working out extensions always subtract the original unstretched length from the loaded length. It is easy to think that the extension is just the extra length after adding one more mass to the hanger.

(d) Explain how the student can calculate the Young modulus from the graph of stress against strain. **(3 marks)**

From the equation $\sigma = E\varepsilon$, a graph of stress σ (y-axis) against strain ε (x-axis) will be a straight line if the wire obeys Hooke's law. In this linear region the Young modulus E is the gradient of the graph, so the student must calculate the gradient.

If you are doing this experimentally don't forget to include units. The gradient is stress divided by strain, so the units are N m^{-2} (strain is dimensionless).

Newton's laws of motion

Three laws define force and explain motion – amazing!

Newton's first law of motion

A body will remain at rest or continue to move in a straight line at a constant velocity unless an external force (an unbalanced or resultant force) acts on it.

The Voyager spacecraft is now moving through deep space. There are virtually no forces acting on the craft so it continues to move at almost constant velocity of about 15 km s⁻¹.

F ←   → D

A car continues to travel at constant velocity along a road when the driving force and drag are equal and opposite. The resultant force on the car is zero.

Newton's second law of motion

The resultant force on an object is proportional to the rate of change of momentum of the object, and the momentum change takes place in the direction of the force.

Momentum, p, is mass × velocity, so when a force is applied to an object that has an unchanging mass, the acceleration is directly proportional to the force.

$$p = mv$$
$$F \propto \frac{\Delta p}{\Delta t} \propto \frac{m\Delta v}{\Delta t} = ma$$
$$F = ma$$

The thrust, T, on this missile is greater than the weight, W. The resultant force is $F = T - W$. The missile accelerates upward. $a = \frac{(T - W)}{m}$, where m is the mass of the missile.

Worked example

(a) The Earth's gravity exerts an attractive force on the Moon, but the Moon orbits the Earth at an almost constant speed. Does this contradict Newton's first law? **(2 marks)**

No. The direction of the Moon's velocity is continually changing so it is accelerating towards the Earth. However, its momentum keeps it moving in a circular orbit.

(b) A car has a driving force of 8000 N and a mass of 1200 kg. The total drag on the car is 6200 N. Calculate its acceleration. **(3 marks)**

Resultant force $F = 8000 - 6200 = 1800$ N.
$a = \frac{F}{m} = \frac{1800}{1200} = 1.5\,\text{m s}^{-2}$

(c) A footballer kicks a stationary ball. The ball accelerates rapidly and then moves towards the goal at almost constant speed.
If action and reaction are equal why don't they cancel out? **(2 marks)**

They are equal but they act on different bodies so they cannot cancel or balance one another. In this case the foot exerts a force on the ball and the ball exerts an equal but opposite force back on the foot.

Newton's third law of motion

If object A exerts a force on object B, then B will exert an equal force in the opposite direction on A.

These pairs of forces are often called action and reaction. They are the same type of force (e.g. contact forces or gravitational forces) and they must act on different bodies.

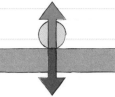

contact force from surface on ball

contact force from ball on surface

Look at the diagram of a ball resting on the ground.

✗ The downward force is **not** the weight of the ball – it is an electrostatic contact force.

✓ The weight of the ball is a gravitational force. It is equal and opposite to the gravitational pull of the ball on the Earth.

Now try this

When a ball bearing is released at the top of a measuring cylinder containing a viscous liquid, such as glycerol, the ball bearing accelerates at first and then reaches a constant terminal velocity. Explain this by referring to Newton's three laws of motion. **(6 marks)**

Linear momentum

The concept of momentum helps us to understand collisions and interactions.

Linear momentum

Linear momentum = mass × velocity

$$p = mv$$

The S.I. unit is $kg\,m\,s^{-1}$ or $N\,s$.

Linear momentum is a vector quantity – its magnitude is mv and its direction is parallel to the velocity vector.

We often simply refer to 'momentum' – this is always **linear** momentum.

25 m s⁻¹

6.0 kg

The object above has a momentum of

$$6.0 × 25 = 150\ kg\,m\,s^{-1}\ \text{to the right.}$$

Force and rate of change of momentum

$F →$ m

A resultant force makes something accelerate. This changes its momentum.

$$F = ma = \frac{m(v - u)}{t}$$

This can be written differently:

$$F = \frac{mv - mu}{t}$$

$$\text{Force} = \frac{\text{change of momentum}}{\text{time}}$$

Resultant force = rate of change of linear momentum

$$F = \frac{\Delta p}{\Delta t}$$

Total momentum of a system

You have to add momenta as vectors.

Whilst each of the rugby players above had a great deal of linear momentum before they collided, the sum is almost zero because they were running in opposite directions. After the collision they both stop moving.

Impacts

The momentum of the hammer is reduced rapidly to zero when it hits the nail. This large rate of change of momentum means a large force is exerted on the hammer and on the nail, driving it in.

$$F = \frac{\Delta p}{\Delta t}$$

Bending your knees on landing increases the time (Δt) over which your momentum falls to zero. This reduces the rate of change of momentum and so reduces the force on your body.

Worked example

A dart of mass 30 g is flying towards a dartboard with a horizontal velocity of $4.0\ m\,s^{-1}$. When it strikes the board it stops in a time of 2.5 ms.

(a) Calculate the momentum of the dart just before it hits the dartboard. **(2 marks)**

$$p = mv = 0.030 × 4.0 = 0.12\ kg\,m\,s^{-1}$$

(b) Calculate the average force on the dart as it comes to rest after hitting the dartboard. **(2 marks)**

$$F = \frac{\Delta p}{\Delta t} = \frac{0.12}{0.0025} = 48\ N$$

Now try this

1 A charging elephant of mass 3000 kg is running at $10\ m\,s^{-1}$. How fast would a car of mass 1200 kg have to move to have the same linear momentum as the elephant? **(2 marks)**

2 Explain why rubberised surfaces are often put down under swings in children's playgrounds. **(3 marks)**

Impulse

Impulses change momentum.

Impulse

Newton's second law of motion can be expressed as an equation.

$$F = \frac{\Delta p}{\Delta t}$$

If this is rearranged:

$$F\Delta t = \Delta p = mv - mu$$

Force × time ($F\Delta t$) is called impulse and is equal to the change of linear momentum, Δp.

The S.I. unit for impulse is the same as that of momentum, $kg\,m\,s^{-1}$ or $N\,s$.

Crumple zones

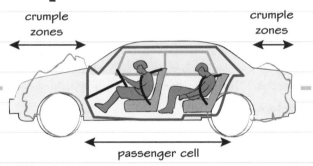

crumple zones — crumple zones

passenger cell

During a collision, people in a car undergo a rapid change of momentum. This means they experience a large impulse. Crumple zones around the car increase the time in which they come to rest. This reduces the force applied to the people.

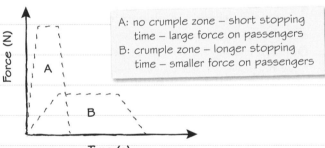

A: no crumple zone – short stopping time – large force on passengers
B: crumple zone – longer stopping time – smaller force on passengers

The areas of A and B are equal; the impulse and change of momentum are the same in both cases.

Using force-time graphs

Impulse, $F\Delta t$, can be calculated from the area under a graph of force against time.

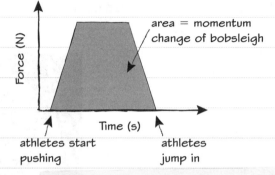

Force (N)

area = momentum change of bobsleigh

Time (s)

athletes start pushing

athletes jump in

This graph shows how the force applied to a bobsleigh varies with time at the start of a run.

Sometimes you will come across an irregularly shaped graph. Remind yourself about estimating areas under graphs on page 10.

Worked example

A ball of mass 0.20 kg falls vertically and hits the ground at a speed of $4.2\,m\,s^{-1}$. It rebounds vertically at an initial speed of $3.4\,m\,s^{-1}$.

(a) Calculate the change of momentum of the ball. **(3 marks)**

Taking downward as positive

$mu = 0.20 \times 4.2 = 0.84\,kg\,m\,s^{-1}$

$mv = -0.20 \times 3.4 = -0.68\,kg\,m\,s^{-1}$

$\Delta p = mv - mu = -0.68 - 0.84$
$= -1.52\,kg\,m\,s^{-1}$

(b) What is the impulse applied to the ball by the ground? **(2 marks)**

Impulse = change of momentum = $-1.52\,N\,s$ or $1.52\,N\,s$ upward.

(c) The average force exerted on the ball while it was in contact with the ground was 18 N. Calculate the contact time. **(2 marks)**

Impulse = $F\Delta t$ so $\Delta t = \frac{1.52}{18} = 0.08\,s$

Now try this

A child purchases a model rocket with a motor claimed to burn for 2.0 s and provide an impulse of 5.0 N s. The rocket and its motor have a combined mass of 100 g.

(a) What change of momentum can the motor provide for the model rocket? **(1 mark)**

(b) What is the average force applied to the rocket by the motor? **(1 mark)**

(c) Calculate a value for the maximum velocity of the rocket based on the impulse provided by the motor. **(2 marks)**

(d) Give two reasons why the actual maximum velocity of the rocket is likely to be different from the value you have calculated in (c). **(2 marks)**

Conservation of linear momentum – collisions in one dimension

The law of conservation of momentum lets you predict what will happen after a collision, an explosion or other interaction.

Total momentum does not change in an interaction

When two objects interact they exert forces on one another. These forces are equal and opposite (Newton's third law). Since they act on each object for the same time, they provide equal and opposite impulses, which cause equal and opposite changes of momentum. The total momentum of the two bodies is the same before and after the interaction. Momentum is always conserved.

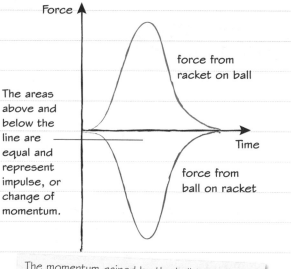

Force

The areas above and below the line are equal and represent impulse, or change of momentum.

force from racket on ball

Time

force from ball on racket

The momentum gained by the ball to the left is lost by the racket. The momentum of the system (ball plus racket) has not changed.

The law of conservation of linear momentum

The linear momentum of a closed system is constant.

A closed system has no external resultant force acting on it.

Collisions in one dimension

If no external forces act on the system the total momentum before and after the collision is the same.

How to solve momentum problems:

1 Write down the linear momentum before the collision.

2 Write down the linear momentum after the collision.

3 Equate the two expressions and solve them for any unknown quantity.

Don't forget to take direction into account!

Here is a collision between two objects that stick together. The object on the right is initially stationary.

Before: After:

$u_1 \longrightarrow$ $u_2 = 0$ $v \longrightarrow$

m_1 m_2 $m_1 + m_2$

Momentum before $= m_1u_1 + m_2u_2 = m_1u_1 + 0$

Momentum after $= (m_1 + m_2)v$

Therefore $(m_1 + m_2)v = m_1u$

$$v = \frac{m_1u}{(m_1 + m_2)}$$

In examples where the objects move in different directions, define one direction as positive. Objects moving in the opposite direction then have negative values of momentum.

If the objects do not stick together after the collision then the expression for final momentum has two terms: $m_1v_1 + m_2v_2$.

Worked example

A car of mass 1400 kg travelling at 10 m s⁻¹ collides with a stationary van of mass 2500 kg.
The two vehicles lock together after the collision.
Calculate the initial velocity of the two vehicles after the collision.   **(3 marks)**

Momentum before $= 1400 \times 10 = 14\,000\,kg\,m\,s^{-1}$

Momentum after $= (1400 + 2500)v$

$$v = \frac{14\,000}{(1400 + 2500)} = 3.6\,m\,s^{-1}$$

Now try this

A woman of mass 60 kg jumps out of a stationary rowing boat onto the bank of a river. She jumps forward at a velocity of 2.0 m s⁻¹. The boat moves backwards at 0.40 m s⁻¹. What is the mass of the boat?   **(3 marks)**

Collisions in two dimensions

Every component of linear momentum is conserved separately.

Two-dimensional collisions

To solve two-dimensional collision problems (A level only):

 1   choose two perpendicular directions with the origin at the position of impact

2   resolve the momentum along each direction

3   use conservation of momentum along each axis to solve problems.

The diagram shows a collision between two balls with masses m_A (moving) and m_B (stationary).

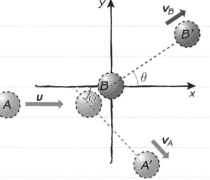

- In the x-direction: $m_A u = m_A v_A \cos\phi + m_B v_B \cos\theta$
- In the y-direction: $0 = m_A v_A \sin\phi - m_B v_B \sin\theta$

By selecting the x-axis parallel to the direction of motion of the incoming ball, the problem is simplified (there is no initial y-component of momentum).

> Resolving vectors into two perpendicular components is covered on page 8.

Using a vector triangle

Momentum is conserved, so the sum of momentum vectors before a collision must equal the sum after the collision. This can be drawn as a vector diagram:

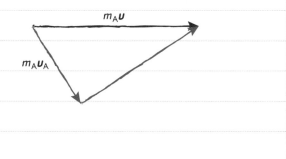

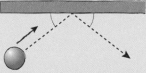 **Worked example**

A football of mass m is kicked against a wall. It approaches the wall at speed v and rebounds at the same speed as shown below.

Use the law of conservation of momentum to explain why the ball rebounds at the same angle. **(3 marks)**

Each component of momentum is conserved. The component parallel to the surface of the wall is $mv\cos\theta$. m and v are unchanged by the collision, so θ must be the same too.

Now try this

A car of mass 1400 kg is travelling at a speed of 25 m s⁻¹ along a road on a bearing of 20° (20° east of north).

(a) Calculate the component of its linear momentum due north. **(2 marks)**

(b) Calculate the component of its linear momentum due east. **(2 marks)**

Elastic and inelastic collisions

Momentum is conserved in all collisions. In some collisions, kinetic energy is conserved too.

Collision types

Elastic collision	Inelastic collision	
momentum is conserved	momentum is conserved	Momentum is always conserved.
kinetic energy is conserved – no energy is transferred to other forms such as sound	some kinetic energy is transferred to other forms of energy	
linear momentum $p = m_1v_1 = m_2v_2$ $E_K = \frac{1}{2}m_1v_1^2 = \frac{1}{2}m_2v_2^2$	linear momentum $p = m_1v_1 = m_2v_2$ E_K changes	

Worked example

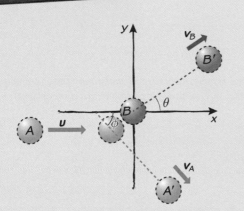

Look again at the collision of two balls.

(a) If this is an elastic collision, write an equation for the conservation of kinetic energy in this collision. **(2 marks)**

$E_{K\ before} = E_{K\ after}$
$\frac{1}{2}mu^2 = \frac{1}{2}mv_A^2 + \frac{1}{2}mv_B^2$

(b) In fact, the collision is not perfectly elastic. Suggest how the kinetic energy is transferred. **(2 marks)**

Most of the initial kinetic energy of A will be transferred into kinetic energy of A and B after the collision, but some will be transferred into sound as the balls collide and by heating through friction. This is an inelastic collision.

Worked example

A football of mass 430 g travelling at 3.5 m s⁻¹, collides with a smaller ball of mass 28 g travelling in the opposite direction at 1.8 m s⁻¹. The velocity of the football after the collision is 3.0 m s⁻¹ in the original direction.

(a) Calculate the new velocity of the smaller ball. **(4 marks)**

Total linear momentum before collision

$p = m_1v_1 + m_2v_2$
$= 0.430 \times 3.5 + 0.028 \times -1.8$
$= 1.4546\ kg\,m\,s^{-1}$

p before collision = p after collision
$1.4546 = m_1'v_1' + m_2'v_2'$

$v_2' = \dfrac{(1.4546 - m_1'v_1')}{m_2'}$

$= \dfrac{(1.4546 - 0.430 \times 3.0)}{0.028}$

$= 5.9\ m\,s^{-1}$ in the direction of the football

(b) Calculate the total kinetic energy before and after the collision, and state whether the collision was elastic or inelastic. **(3 marks)**

Total E_K before = $\frac{1}{2}m_1v_1^2 + \frac{1}{2}m_2v_2^2$
$= 0.5 \times (0.430 \times 3.5^2 + 0.028 \times 1.8^2)$
$= 2.7\ J$

Total E_K after = $\frac{1}{2}m_1'v_1'^2 + \frac{1}{2}m_2'v_2'^2$
$= 0.5 \times (0.430 \times 3.0^2 + 0.028 \times 5.9^2)$
$= 2.4\ J$

The collision is inelastic because E_K is not conserved.

Now try this

A spring-loaded toy cannon of mass 213.0 g fires table-tennis balls of mass 2.7 g. On one firing, after the initial impulse the cannon recoiled with a velocity of 35 cm s⁻¹.

(a) Calculate the initial velocity of the table-tennis ball. **(2 marks)**

(b) Calculate the kinetic energy before and after firing, and explain why it is not conserved. **(2 marks)**

(c) Comment on whether the toy is dangerous and explain your reasoning. **(2 marks)**

Electric charge and current

Electrical effects can be explained in terms of the movement of charge, usually tiny charged particles called electrons.

What is charge?

- Charge is a fundamental property of some subatomic particles.
- Charges have electric fields around them.
- Charges exert forces on other charges.
- Charge is quantised – the smallest unit of charge is the charge, e, on an electron, and charge always occurs in multiples of e.
- Like charges repel one another.
- Opposite charges attract one another.

Measuring charge

The S.I. unit of charge is the coulomb (C).

The smallest free charge is the charge on an electron: $e = -1.6 \times 10^{-19}$ C.

Charge can be measured directly using a coulombmeter.

If a charge is transferred to the metal cap the meter gives a reading in nC (nanocoulombs).

The electron

- Electrons are all identical.
- Electrons are present in all atoms.
- Electrons carry a negative charge.
- Protons carry a positive charge.
- Electrons and protons carry the same size of charge.
- When an atom loses electrons, it becomes a positive **ion**.
- When an atom gains electrons, it becomes a negative ion.
- Ions can carry charge when they are free to move, for example in liquid **electrolytes**.

Current and charge

Electric **current** is the rate of flow of electric charge.

$$\text{current} = \frac{\text{charge flow}}{\text{time}}$$
$$I = \frac{\Delta Q}{\Delta t}$$

The S.I. unit of current is the ampere, usually shortened to amp (A). A current of 1 A is a flow of one coulomb per second, $1\,\text{C s}^{-1}$.

Current is measured using an ammeter.

Ammeters must be connected in **series**.

Current in a circuit

By convention, current direction is said to be from positive to negative around a circuit:

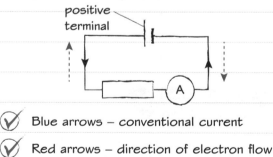

positive terminal

- ✓ Blue arrows – conventional current
- ✓ Red arrows – direction of electron flow

Worked example

When a car battery is used to turn the starter motor of a car, a current of 120 A flows for 3.0 s.

(a) How much charge passes through the motor?

(2 marks)

$\Delta Q = I\Delta t = 120 \times 3.0 = 360\,\text{C}$

(b) The battery is rated at 85 A h, which means that in principle it can supply a constant current of 1.0 A for 85 hours. How much charge passes through the battery before it needs recharging?

(2 marks)

$\Delta Q = I\Delta t = 85 \times 60 \times 60 \times 1.0$
$\qquad = 306\,000\,\text{C}$

(c) Sometimes a car will not start because it has a 'flat battery'. Does this mean that the battery has 'run out of charge'?

(3 marks)

No. A battery transfers chemical energy to electrical energy to make charge flow in the circuit, but it does not supply the charge. It has run out of the chemical energy source that allows it to do this.

Now try this

1 A camera flash gun works by storing charge on a capacitor and then letting the capacitor discharge rapidly. The charge stored is $10\,\mu\text{C}$ and it is discharged in 1.0 ms. Calculate the average current that flows while it discharges. **(3 marks)**

2 A lightning strike carries a current of 50 kA and transfers 1.25 C of charge between the storm cloud and the ground. How long does the current flow? **(3 marks)**

Charge flow in conductors

Current is rate of flow of charge, but the actual charges that flow differ from one conductor to another.

Conductors and insulators

Conductors are materials that allow charge to flow through them, like metals and electrolytes.

Insulators are materials that do not allow charge to flow through them, like glass and rubber.

Electrical breakdown: If the **potential difference** across an insulator is increased, eventually it might be high enough to cause the insulator to begin to conduct.

Charge carriers are the charged particles that carry electric charge by moving through a conductor, often electrons or ions.

Metallic conductors

Metals have a crystalline structure in which some electrons are delocalised (shared). These **free electrons** can move easily through the metal, so metals are good electrical conductors.

In a metal the electrons drift in the opposite direction to the conventional current.

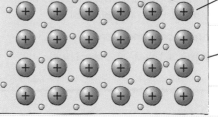

regular lattice of metal ions

Free electrons can drift through the metal when a potential difference is applied.

Electrolytes

Electrolytes contain ions in solution.

✓ Positive ions flow in the direction of conventional current.

✓ Electrons and negative ions flow in opposite direction to conventional current.

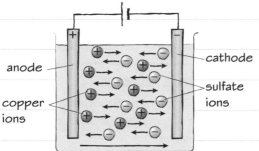

anode

copper ions

cathode

sulfate ions

direction of conventional current

Conduction in copper sulfate solution

Conduction through the air

If the potential difference between two conductors with an air gap between them is increased enough, regions of air in the gap undergo electrical breakdown. Molecules in the air are ionised by the electric field and an avalanche of charge flows. This is a spark.

Lightning is an extreme example of electrical breakdown in the air.

Semiconductors

Some materials called **semiconductors** have a small number of free electrons at room temperature, so they do just conduct but they have much higher resistivity than a metal. However, unlike metals, their resistivity falls as temperature rises.

Resistivity is covered on pages 48 and 49.

Silicon and germanium are semiconductors.

Worked example

A current of 40 mA flows between two copper electrodes immersed in copper sulfate solution. The current in the solution is carried by Cu^{2+} ions and SO_4^{2-} ions.

(a) Calculate the charge that is transferred in 5.0 minutes. **(2 marks)**

$\Delta Q = I\Delta t = 0.040 \times 5.0 \times 60 = 12\,C$

(b) Explain why copper is deposited on the cathode (negative electrode). **(2 marks)**

Copper ions are positively charged so they are attracted to the negative electrode (cathode) and discharged. When a copper ion is discharged it becomes a copper atom and is deposited on the electrode.

Now try this

Look at the worked example.

(a) How many copper ions are deposited on the cathode in 5.0 minutes? ($e = 1.60 \times 10^{-19}\,C$) **(2 marks)**

(b) How long would it take to deposit 1.0 g of copper onto the cathode? (The mass of 1 mole of copper atoms is 63.5 g and the Avogadro constant is $6.02 \times 10^{23}\,mol^{-1}$.) **(5 marks)**

Kirchhoff's first law

How does current behave at a junction? Like traffic, some charge goes one way and some goes the other way!

Conservation of electric charge

The total amount of charge in a closed system is constant.

Electric circuits are closed systems because charge cannot escape from the conductors; it can only move around the circuit. So the total amount of electric charge in a circuit is constant.

Batteries

A cell or battery cannot run out of electrical charge – it runs out of energy. Rechargeable cells can have their internal stores of chemical energy topped up.

Kirchhoff's first law for electric circuits

The sum of currents entering a junction in an electric circuit is equal to the sum of currents leaving that junction.

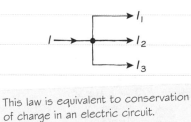

This law is equivalent to conservation of charge in an electric circuit.

Current in a parallel circuit

In a **parallel** circuit, charge can follow more than one route.

The current splits at each junction.
The amount of current in each branch depends on the resistance in the branch and the potential difference across it.

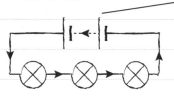

Current recombines at each junction. The current through the battery is equal to the sum of currents through each branch.

Current in a series circuit

A **series** circuit has components arranged one after the other. There are no junctions in a series circuit, so the charge can flow through only one route. The current at all points in a series circuit is the same.

The charge also passes through the battery.

Worked example

The diagram below shows part of an electric circuit. The currents at two points are shown.

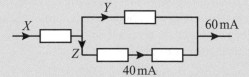

What are the currents X, Y and Z?   (3 marks)

Applying Kirchhoff's first law

  60 mA = 40 mA + Y, so Y = 20 mA

  Z = 40 mA (in series with the 40 mA current).

  X = Y + Z = 60 mA

Now try this

The diagram shows currents at a junction in an electric circuit. Calculate the value of I_2 and state its direction of flow.   (3 marks)

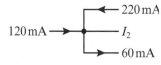

Charge carriers and current

In order to explain current on the large scale, we need to understand the processes that allow charge to flow at the atomic scale.

A microscopic equation for current

Consider a steady current, I, in a conductor like a wire. Inside the conductor, current is the flow of microscopic charge carriers. These might be positively charged (e.g. ions) or negatively charged (e.g. electrons).

Each electron has a charge $-e$ and a mean drift velocity v. There are n electrons per unit volume V.

In time Δt the charge carriers move forward a distance $v\Delta t$.

The volume through which charge carriers pass in time Δt is $Av\Delta t$.

The total charge passing a point in Δt is therefore $\Delta Q = nAve\Delta t$.

Current $I = \dfrac{\Delta Q}{\Delta t} = Anev$

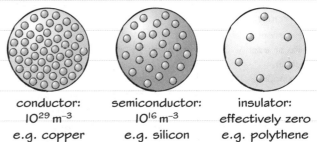

charge carrier (electron)　　cylindrical conductor

conventional current, I

cross-sectional area, A

Drift velocity

Charge carriers do not all move with a constant velocity in a straight line. They are continually scattered from atoms and ions in the material. This transfers energy by heating.

Drift velocities in metals are surprisingly low (look at the worked example). The drift velocity in semiconductors is much higher because the density of charge carriers (n) is much lower than in a metal.

Conductors, semiconductors and insulators

conductor:
10^{29} m^{-3}
e.g. copper

semiconductor:
10^{16} m^{-3}
e.g. silicon

insulator:
effectively zero
e.g. polythene

Charge carrier densities for different materials

Charge carrier density in semiconductors increases with temperature and can be changed by 'doping' the semiconductor with atoms with a different valency.

Worked example

A copper wire of diameter 0.32 mm carries a steady current of 100 mA.

The charge carrier density in copper is 8.5×10^{28} m^{-3}. The charge on an electron is 1.60×10^{-19} C.

(a) Calculate the number of electrons that pass a point in the wire every minute. **(3 marks)**

In one minute $\Delta Q = I\Delta t = 0.10 \times 60 = 6.0$ C

Number $= \dfrac{6.0}{1.60 \times 10^{-19}} = 3.75 \times 10^{19}$

(b) Calculate the mean drift velocity of these electrons. **(4 marks)**

$I = Anev$, so $v = \dfrac{I}{Ane} = \dfrac{I}{\pi\left(\dfrac{d}{2}\right)^2 ne}$

$v = \dfrac{0.10}{\pi\left(\dfrac{0.32}{2} \times 10^{-3}\right)^2 \times 8.5 \times 10^{28} \times 1.60 \times 10^{-19}}$

$= 9.1 \times 10^{-5}$ m s^{-1}

Now try this

A simple circuit lights a filament lamp from a cell. The connecting wires are copper and have a diameter of 0.25 mm and a total length of 50.0 cm. When operating at normal brightness the lamp draws a current of 250 mA.

(a) Calculate the number of electrons per second that leave the cell. **(2 marks)**

(b) Calculate the mean drift velocity of these electrons in the copper connecting wires. **(4 marks)**

(c) Calculate the time taken for one electron to pass once around the circuit (a distance of 50 cm). **(2 marks)**

(d) Explain why the lamp comes on as soon as the circuit is connected despite the slow drift velocity of the electrons. **(2 marks)**

Electromotive force & potential difference

Energy is transferred in electric circuits – e.m.f. and p.d. are related to the energy transfers.

Energy transfers in a simple electrical circuit

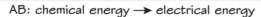

AB: chemical energy → electrical energy

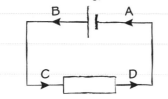

CD: electrical energy → heat energy

The cell is a **source** of electrical energy and has an **electromotive force** (e.m.f.) across it.

The resistor transfers electrical energy to other forms and has a **potential difference** (p.d.) across it.

Using voltmeters

Voltmeters must be connected in parallel because they are measuring the potential **difference** between two points in a circuit.

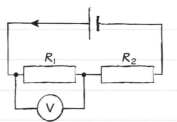

The voltmeter above measures the p.d. across R_1.

An ideal voltmeter should have infinite resistance so that it does not affect the circuit it is connected to.

p.d. and energy

When a charge Q moves through a p.d. V, the work done, W, by the charge is equal to QV.

$$W = QV$$

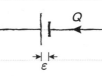

When a charge, Q, moves through an e.m.f., ε, the work done, W, on the charge is equal to εQ.

$$W = \varepsilon Q$$

Since $Q = It$, the equation $W = \varepsilon It$ or $W = VIt$ can be used to calculate energy transfers over a period of time during which a constant current flows.

When a p.d. V is applied across a conductor, the charge carriers are accelerated and gain kinetic energy equivalent to eV

$$\tfrac{1}{2}mv^2 = eV$$

e.m.f. and p.d.

e.m.f. – the energy gained per unit charge by charges passing through the supply, when a form of energy is transferred to electrical energy carried by the charges.

p.d. across a component – the energy transferred per unit charge by the charges passing through the component.

The unit for both of these is the volt (V) or joule per coulomb ($J\,C^{-1}$).

Sometimes people refer to e.m.f. and p.d. as 'voltage'.

Definition of the volt

The p.d. (or e.m.f.) between two points in an electric circuit is 1 V if 1 J of electrical energy is transferred for every 1 C of charge that moves between those two points.

Worked example

Look at the circuit on the right.

(a) How much charge passes through R_1 in 1.0 s?
(2 marks)

$Q = It = 2.0 \times 1.0$
$= 2.0\,C$

(b) How much energy is transferred by R_1 in 1.0 s?
(2 marks)

$W = QV = 2.0 \times 8.5 = 17\,J$

(c) How much energy is transferred by R_2 in 1.0 s?
(2 marks)

$W = QV = 2.0 \times 3.5 = 7.0\,J$

(d) What is the e.m.f. of the cell? Explain your answer. **(3 marks)**

$W = \varepsilon Q$ so $\varepsilon = \dfrac{W}{Q} = \dfrac{17 + 7.0}{2.0} = 12\,V$.

The energy used by the two resistors must equal the energy supplied by the cell.

Now try this

An AA cell has an e.m.f. of 1.5 V.

(a) How much energy does it supply to an electric circuit per unit of charge that passes through it?
(1 mark)

(b) When the AA cell is used in a simple circuit it supplies a steady current of 200 mA for 5.0 minutes. How much energy does it supply? **(3 marks)**

Resistance and Ohm's law

The relationship between the p.d. across a component and the current through it defines the component's electrical resistance.

Definition of resistance

The **resistance**, R, of a component is the ratio of the potential difference across the component to the current through it.

$$R = \frac{V}{I}$$

The S.I. unit of resistance is the ohm (Ω).
$1\,\Omega = 1\,V\,A^{-1}$

Measuring resistance with an ohmmeter

A multimeter can be used as an ohmmeter by connecting to the common and Ω terminals and selecting a resistance range.

Ohmmeters must only be used to measure the resistance of isolated components.

Do not use ohmmeters in a working circuit!

Ohm's law

Current through a conductor is directly proportional to p.d., provided that physical conditions, such as temperature, remain constant.

$I \propto V$ or $\dfrac{V}{I}$ is constant.

Not all conductors obey Ohm's law. Those that do are described as ohmic conductors.

Measuring resistance with an ammeter and voltmeter

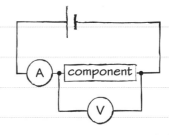

Calculate $R = \dfrac{V}{I}$. For an ohmic conductor, R is equal to the reciprocal of the gradient of a graph of I against V.

Worked example

The table shows the current through a filament lamp when it is connected to two different power supplies.

	P.d. (V)	Current (A)
Power supply A	12.0	2.0
Power supply B	13.0	2.05

(a) Calculate the resistance of the lamp when connected to supply A and then to supply B.

(2 marks)

$R_A = \dfrac{V}{I} = \dfrac{12.0}{2.00} = 6.0\,\Omega$

$R_B = \dfrac{V}{I} = \dfrac{13.0}{2.05} = 6.3\,\Omega$

(b) Is the material in the lamp filament an ohmic conductor? Explain your answer. **(3 marks)**

No. The resistance has increased so V/I is not constant. A graph of I against V would not be a straight line through the origin so I is not directly proportional to V.

Now try this

Copy and complete the table for a number of different conductors. **(4 marks)**

Component	P.d.	Current	Resistance
W	40 V		200 Ω
X	6.0 V	50 mA	
Y		30 mA	20 kΩ
Z	3.0 kV	6.0 μA	

Take care over the units! It is usually best to convert everything to S.I. base units (A, V, Ω) before carrying out the calculation.

I–V characteristics

Graphs of current against p.d. show how different electrical components behave in an electric circuit.

An ohmic conductor (constant resistance at constant temperature)

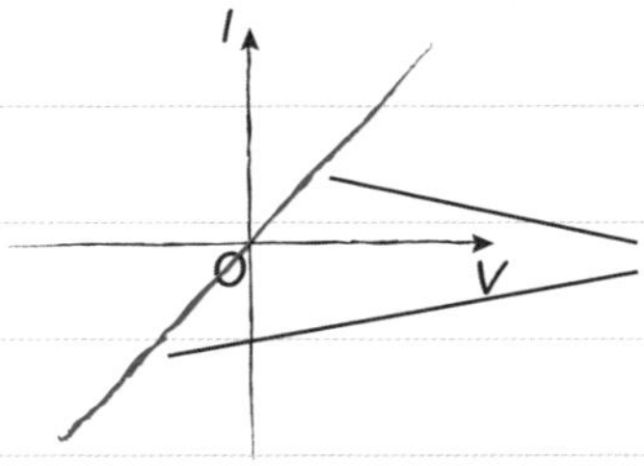
I
O
V

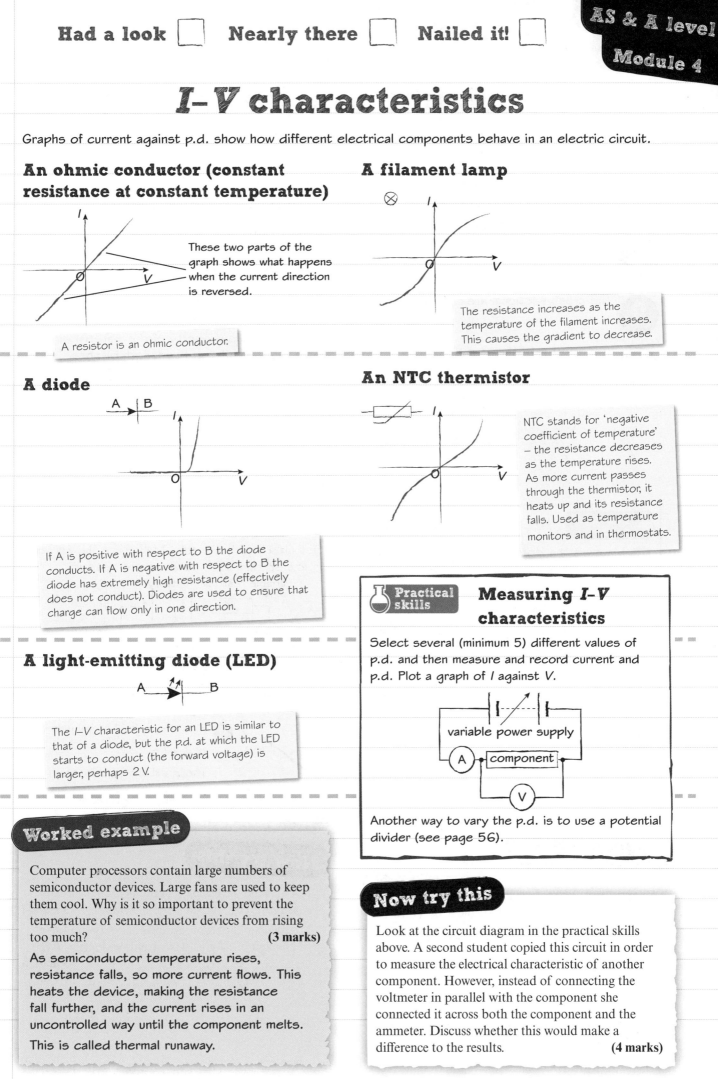

These two parts of the graph shows what happens when the current direction is reversed.

A resistor is an ohmic conductor.

A filament lamp

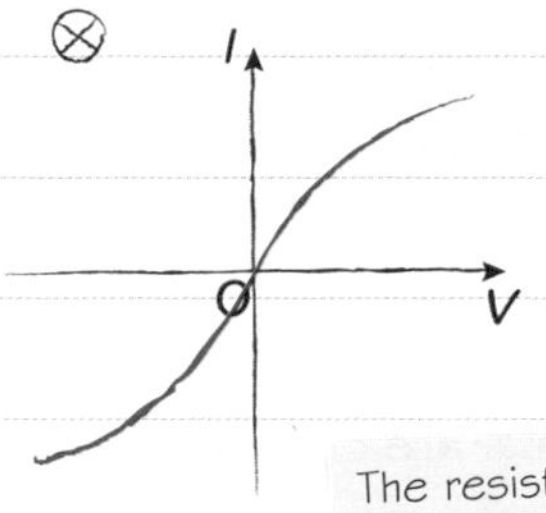
I
O
V

The resistance increases as the temperature of the filament increases. This causes the gradient to decrease.

A diode

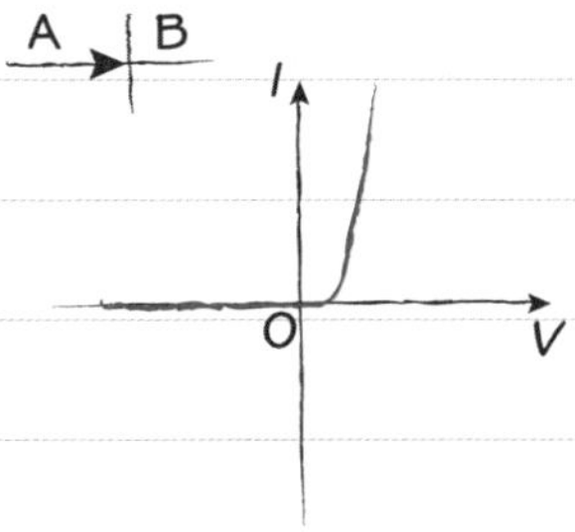
A
B
I
O
V

If A is positive with respect to B the diode conducts. If A is negative with respect to B the diode has extremely high resistance (effectively does not conduct). Diodes are used to ensure that charge can flow only in one direction.

An NTC thermistor

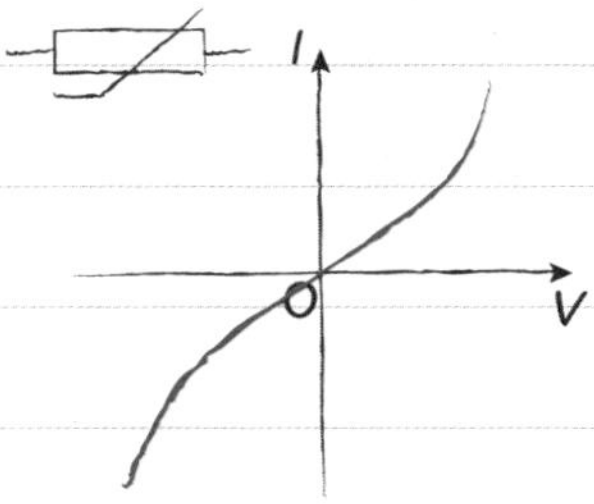
I
O
V

NTC stands for 'negative coefficient of temperature' – the resistance decreases as the temperature rises. As more current passes through the thermistor, it heats up and its resistance falls. Used as temperature monitors and in thermostats.

A light-emitting diode (LED)

A
B

The *I–V* characteristic for an LED is similar to that of a diode, but the p.d. at which the LED starts to conduct (the forward voltage) is larger, perhaps 2 V.

Practical skills: Measuring *I–V* characteristics

Select several (minimum 5) different values of p.d. and then measure and record current and p.d. Plot a graph of *I* against *V*.

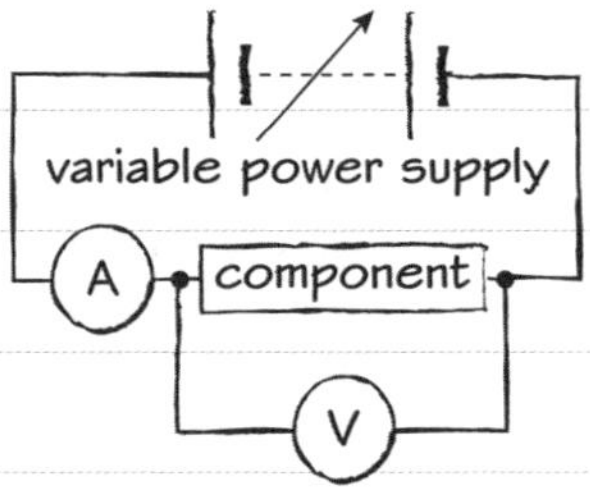
variable power supply
A
component
V

Another way to vary the p.d. is to use a potential divider (see page 56).

Worked example

Computer processors contain large numbers of semiconductor devices. Large fans are used to keep them cool. Why is it so important to prevent the temperature of semiconductor devices from rising too much? **(3 marks)**

As semiconductor temperature rises, resistance falls, so more current flows. This heats the device, making the resistance fall further, and the current rises in an uncontrolled way until the component melts.

This is called thermal runaway.

Now try this

Look at the circuit diagram in the practical skills above. A second student copied this circuit in order to measure the electrical characteristic of another component. However, instead of connecting the voltmeter in parallel with the component she connected it across both the component and the ammeter. Discuss whether this would make a difference to the results. **(4 marks)**

Resistance and resistivity

Resistance is the property of a component, like mass; resistivity is the property of a material, like density.

Resistance

The resistance of a piece of wire at constant temperature depends on three things:

 1 the length of the wire

 2 the cross-sectional area of the wire

3 the material of the wire.

$R \propto l$　　$R \propto \dfrac{1}{A}$

Resistivity

The constant is called the resistivity, ρ.

$$R = \text{constant} \times \frac{l}{A} \text{ (see left)}$$

$$R = \frac{\rho l}{A}$$

S.I. units for resistivity are $\Omega\,m$.

Typical resistivity values

Resistivity is a property of the material of the wire – it is unaffected by length or area.

Material	ρ ($\Omega\,m$)
copper	1.72×10^{-8}
aluminium	2.82×10^{-8}
tungsten	5.60×10^{-8}
nichrome	150×10^{-8}

Practical skills — **Measuring resistivity**

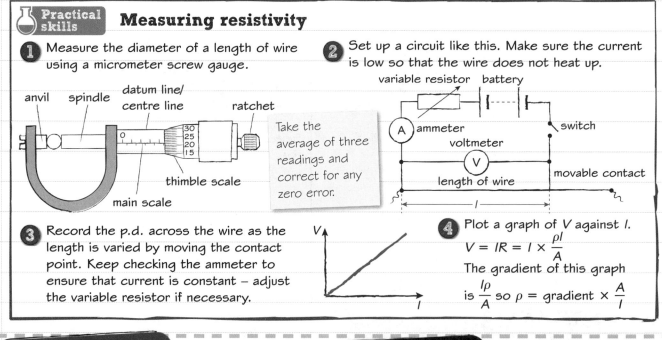

1 Measure the diameter of a length of wire using a micrometer screw gauge.

anvil　spindle　datum line/centre line　ratchet

thimble scale

main scale

Take the average of three readings and correct for any zero error.

2 Set up a circuit like this. Make sure the current is low so that the wire does not heat up.

variable resistor　battery

A ammeter　　switch

voltmeter

V

length of wire　　movable contact

l

3 Record the p.d. across the wire as the length is varied by moving the contact point. Keep checking the ammeter to ensure that current is constant – adjust the variable resistor if necessary.

4 Plot a graph of V against l.
$$V = IR = I \times \frac{\rho l}{A}$$
The gradient of this graph is $\dfrac{I\rho}{A}$ so $\rho = \text{gradient} \times \dfrac{A}{I}$

Worked example

A copper wire 2.00 m long has a resistance of 0.25 Ω. Calculate the diameter of the wire. **(4 marks)**

$$R = \frac{\rho l}{A} \text{ so } A = \frac{\rho l}{R} = 1.72 \times 10^{-8} \times \frac{2.00}{0.25}$$

$$A = 1.376 \times 10^{-7}\ m^2$$

$$A = \pi\left(\frac{d}{2}\right)^2$$

so $d = 2\sqrt{A/\pi} = 2\sqrt{(1.376 \times 10^{-7}/\pi)}$
$$= 4.2 \times 10^{-4}\ m = 0.42\ mm$$

Now try this

1 Calculate the resistance of a nichrome wire of length 20 cm and diameter 0.50 mm. **(3 marks)**

2 Calculate the cross-sectional area of a tungsten filament of length 10.0 cm and resistance 4.0 Ω. **(3 marks)**

Resistivity and temperature

Resistivity is independent of the dimensions of a conductor but it does depend on temperature.

Factors affecting resistivity

The resistivity of a material depends on:

1 the number of charge carriers per unit volume

2 the mobility of the charge carriers. ——— how easily they can move

Both factors can be affected by temperature.

Variation of resistivity with temperature for NTC thermistors

An NTC thermistor is a circuit component made from a semiconductor.

negative temperature coefficient

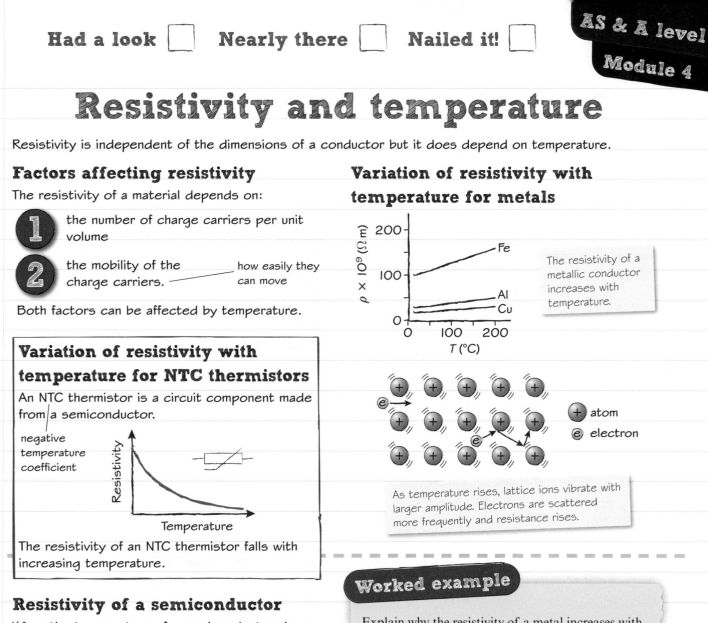

The resistivity of an NTC thermistor falls with increasing temperature.

Variation of resistivity with temperature for metals

The resistivity of a metallic conductor increases with temperature.

As temperature rises, lattice ions vibrate with larger amplitude. Electrons are scattered more frequently and resistance rises.

+ atom
e electron

Resistivity of a semiconductor

When the temperature of a semiconductor rises, more of its charge carriers have enough energy to move. The number of charge carriers per unit volume, n, in $I = Anev$ is increased, so the current increases and the resistivity falls.

Practical skills — Investigating the behaviour of a thermistor

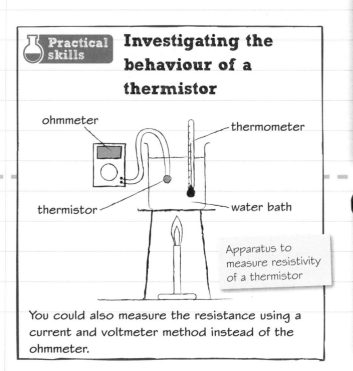

ohmmeter

thermometer

thermistor

water bath

Apparatus to measure resistivity of a thermistor

You could also measure the resistance using a current and voltmeter method instead of the ohmmeter.

Worked example

Explain why the resistivity of a metal increases with temperature but the resistivity of a semiconductor material falls with temperature. **(4 marks)**

In both cases the increase in temperature increases the vibrational energy of particles in the material. This causes more scattering of charge carriers and tends to increase resistivity. In a metal this is the dominant process. In semiconductors there is another process taking place at the same time. As temperature rises, more charge carriers break free of their atoms and are able to move through the material. This process outweighs the extra vibration of the atoms and reduces the resistivity.

Now try this

Look at the graph of resistivity against temperature for metals above.

(a) Describe how the resistivity of aluminium varies with temperature. **(2 marks)**

(b) How does the resistivity of iron differ from the resistivity of copper over the range 50 °C to 150 °C? **(2 marks)**

Electrical energy and power

Electrical energy is a valuable commodity – it is what we pay our electricity bills for.

Electrical energy and electrical power

This circuit transfers electrical energy into heat in a resistor. The p.d. across the resistor is V and there is a steady current I.

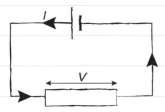

$$\Delta Q = I\Delta t$$

If the current remains constant then the energy transferred in time Δt to the resistor in the circuit above will be

$$\Delta W = V\Delta Q = VI\Delta t$$

Power is the rate of transfer of energy:

$$P = \frac{\Delta W}{\Delta t}$$

where ΔW is the energy transfer that takes place in time Δt.

Electrical power $P = \dfrac{\Delta W}{\Delta t} = \dfrac{VI\Delta t}{\Delta t} = VI$

Power = p.d. × current, $P = VI$

Accelerating charged particles

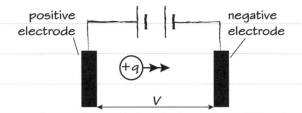

Work done when a charged particle moves through a p.d. V is

$$W = qV$$

If the charge is in a vacuum, electrical energy is transferred entirely to kinetic energy.

$$E_K = \tfrac{1}{2}mv^2 = qV$$

$$v = \sqrt{\frac{2qV}{m}}$$

Power transferred by a resistor

The equation for electrical power can be written in three ways by substituting from $V = IR$.

1 $P = VI$

2 $P = \dfrac{V^2}{R}$ (replacing I in **1** with $\dfrac{V}{R}$)

3 $P = I^2R$ (replacing V in **1** with IR)

Paying for electrical energy

Energy used is calculated in a non-S.I. unit called the kilowatt-hour (kWh). This is the amount of energy transferred by a power of 1 kW in a time of 1 hour.

Energy used in kWh = power (kW) × time (h)

1 kWh = 1000 W × 3600 s = 3.6×10^6 J

Cost = energy used (kWh) × price per kWh

Worked example

An electric shower draws a current of 40 A when operated from a supply of p.d. 230 V.

(a) Calculate the power of the shower. **(2 marks)**

$P = IV = 40 \times 230 = 9200\,\text{W} = 9.2\,\text{kW}$

(b) A typical shower lasts 6.0 minutes. Calculate the energy transferred in this time in kWh and joules. **(3 marks)**

Time taken is 0.10 hours.

Energy (kWh) = 9.2 × 0.10 = 0.92 kWh

Energy (J) = 0.92 × 3.6×10^6 = 3.3×10^6 J

(c) What is the annual cost on the electricity bill if each member of a family of four showers every day and the price of electricity is 20p per kWh? **(3 marks)**

Total energy used = 4 × 365 × 0.92
                              = 1343 kWh

Cost = 1343 × 0.20 = £270 per year.

Now try this

1 An electric kettle is rated at 2.0 kW and operates from a mains voltage (p.d.) of 230 V.
   (a) Calculate the current in the kettle. **(2 marks)**
   (b) Calculate the cost of bringing 1 litre of water to the boil if this takes 5.0 minutes and the cost of electricity is 20p per kWh. **(3 marks)**

2 What p.d. is needed to accelerate an electron from rest to a velocity of $0.05c$ in a vacuum?
   ($c = 3.0 \times 10^8\,\text{m s}^{-1}$, $e = 1.6 \times 10^{-19}\,\text{C}$, $m_e = 9.1 \times 10^{-31}\,\text{kg}$) **(4 marks)**

Kirchhoff's laws and circuit calculations

Kirchhoff's two laws are used to solve circuit problems.

Kirchhoff's first law

Remind yourself about this by looking back at page 43.

The sum of currents entering a junction in an electric circuit is equal to the sum of currents leaving that junction.

$$I = I_1 + I_2 + I_3$$

Kirchhoff's second law

The sum of e.m.f.s is equal to the sum of p.d.s around any closed loop in an electric circuit.

$$\varepsilon = V_1 + V_2$$

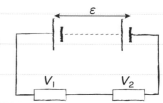

Parallel circuits

Kirchhoff's first law

At A: $I_1 = I_2 + I_3$

At B: $I_2 + I_3 = I_1$

Kirchhoff's second law

In loop 1 (through supply and top branch)

$$\varepsilon = V_1 + V_2$$

In loop 2 (through supply and lower branch)

$$\varepsilon = V_3$$

It is also possible to take a third loop around the two branches without passing through the supply, say clockwise:

$$0 = V_1 + V_2 - V_3$$

The negative sign is because this loop passes through the lower resistor in the opposite direction to the current.

Worked example

In this circuit, the p.d. across the LED when it conducts is about 2 V, and the current through it must be limited to 100 mA.

Calculate the value of the series resistor when the circuit is run from a 6 V supply.   **(4 marks)**

Using Kirchhoff's second law around the circuit:

$$6 = V_R + 2$$

so the p.d. across the series resistor is 4 V.

The current in the circuit is 0.10 A (series circuit).

$$R = \frac{4}{0.10} = 40\,\Omega.$$

Worked example

Use Kirchhoff's laws to find the values of ε, V_1, I_2, I_3, R_2 and R_3 in this circuit.   **(6 marks)**

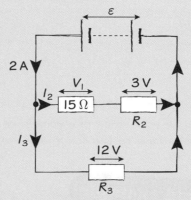

Kirchhoff's second law – outer loop:

$$\varepsilon = 12\,V.$$

Kirchhoff's second law – upper loop:

$$12 = V_1 + 3,\ so\ V_1 = 9\,V.$$

$$I_2 = \frac{V_1}{R} = \frac{9}{15} = 0.6\,A$$

Kirchhoff's first law – left junction:

$$2 = 0.6 + I_3,\ so\ I_3 = 1.4\,A$$

$$R_2 = \frac{V}{I_2} = \frac{3}{0.6} = 5\,\Omega$$

$$R_3 = \frac{V}{I_3} = \frac{12}{1.4} = 8.6\,\Omega$$

Now try this

Calculate the e.m.f., ε, of the battery and the current, I, drawn from the battery in this circuit. All three resistors have a value of 200 Ω.   **(5 marks)**

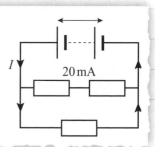

Resistors in series and parallel

The total resistance in a circuit depends on how individual resistors are connected together.

Circuit resistance

When a supply is connected to a circuit containing several components, the circuit resistance is

$$R_{circuit} = \frac{\text{p.d. across circuit}}{\text{current through circuit}}$$

$$R_{circuit} = \frac{V}{I}$$

Resistors in series

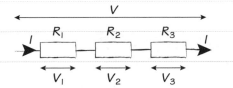

The current through each resistor is I (they are in series).

$$V_1 = IR_1 \qquad V_2 = IR_2 \qquad V_3 = IR_3$$

By Kirchhoff's second law, $V = V_1 + V_2 + V_3$

$$R_{series} = \frac{(V_1 + V_2 + V_3)}{I} = \frac{(IR_1 + IR_2 + IR_3)}{I}$$

$$R_{series} = R_1 + R_2 + R_3$$

So the total resistance of a number of resistors in series is equal to the sum of the individual resistances.

Resistors in parallel

When resistors are connected in parallel, the p.d. across each of them is same.

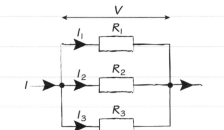

From Kirchhoff's first law, $I = I_1 + I_2 + I_3$

Using $V = IR$ for each resistor

$$\frac{V}{R_{parallel}} = \frac{V}{R_1} + \frac{V}{R_2} + \frac{V}{R_3}$$

$$\frac{1}{R_{parallel}} = \frac{1}{R_1} + \frac{1}{R_2} + \frac{1}{R_3}$$

To find the resistance of a number of resistors in parallel there are two steps.

1 Find the sum of the reciprocals of all the individual resistors.

2 Take the reciprocal of this sum.

Worked example

Calculate the current that is drawn from the supply in this circuit. **(5 marks)**

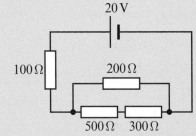

Combining the 500 Ω and 300 Ω resistors in series gives 500 + 300 = 800 Ω.

This 800 Ω is in parallel with 200 Ω.

$$\frac{1}{R_{para}} = \frac{1}{800} + \frac{1}{200} = \frac{5}{800}$$

$$R_{para} = \frac{800}{5} = 160\,\Omega$$

 Remember to convert $\frac{1}{R_{para}}$ into R_{para}!

This 160 Ω is in series with the 100 Ω resistor so the total circuit resistance is

$$100 + 160 = 260\,\Omega$$

The current from the supply is $I = \frac{V}{R} = \frac{20}{260}$
= 0.077 A or 77 mA.

Two resistors

Series: $R_{series} = R_1 + R_2$

Parallel: $\dfrac{1}{R_{parallel}} = \dfrac{1}{R_1} + \dfrac{1}{R_2}$

$$R_{parallel} = \frac{R_1 R_2}{R_1 + R_2}$$

$$R_{parallel} = \frac{\text{product of resistances}}{\text{sum of resistances}}$$

Now try this

Calculate the current drawn from the supply in this circuit. **(5 marks)**

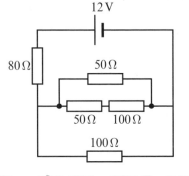

DC circuit analysis

This page shows several worked examples to show how you can solve problems involving direct current (DC) circuits.

Worked example

What is the current through the 50 Ω resistor in the circuit below? **(5 marks)**

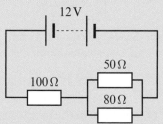

1 Calculate the total resistance.
$$R_{tot} = 100 + \frac{(50 \times 80)}{(50 + 80)} = 131\,\Omega$$

2 Calculate the current drawn from the battery.
$$I = \frac{V}{R} = \frac{12}{131} = 0.092\,A$$

3 Calculate the p.d. across the 100 Ω resistor.
$$V = IR = 0.092 \times 100 = 9.2\,V$$

4 Calculate the p.d. across the parallel resistors.
$$V = 12 - 9.2 = 2.8\,V$$

5 Calculate the current through the 50 Ω resistor.
$$I = \frac{V}{R} = \frac{2.8}{50} = 0.056\,A = 56\,mA.$$

Worked example

The resistors in the circuit below are all of equal resistance. What is the p.d. between points A and B? **(4 marks)**

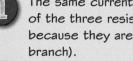

1 The same current flows through each of the three resistors in each branch because they are in series (in the branch).

2 $V = IR$ so the p.d. across each of the resistors is the same, $\frac{12}{3} = 4\,V$.

3 The p.d. between A and the bottom of the circuit as drawn is $4 + 4 = 8\,V$ and the p.d. between B and the bottom of the circuit is 4 V.

4 The p.d. between A and B is $8 - 4 = 4\,V$.

Now try this

1 Calculate the p.d. across the 3.0 kΩ resistor in the circuit below. **(5 marks)**

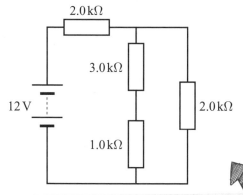

2 The resistance network below consists of five equal 5 Ω resistances. Calculate the current through the cell. **(5 marks)**

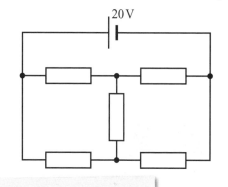

To solve this problem, first find the total resistance of the circuit, then find the current through the cell. Next, find the p.d. across the top 2.0 kΩ resistor and hence the p.d. across the parallel network. Once you have done this you can at last use the potential divider formula to find the p.d. across the 3.0 kΩ resistor.

E.m.f. and internal resistance

Real cells transfer some energy internally when they are used, so not all the energy is transferred in the circuit.

Real cells

Cells transfer chemical energy to electrical energy when a current flows through them. The chemical compounds inside have resistivity, so work has to be done by the charges as they pass through the cell. This work transfers some energy to heat inside the cell and reduces the energy transfer Ir to the external circuit.

As a charge moves through the cell, energy is transferred to it. Some of this energy is immediately transferred to heat inside the cell and the rest is transferred to the external circuit as electrical energy.

A model of a real cell

A real cell can be treated as a perfect source of e.m.f., ε, in series with an **internal resistance**, r.

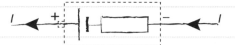

When current flows through the cell there will be a p.d. across the internal resistance equal to Ir.

The **terminal p.d.** of the cell is therefore:

$$V = \varepsilon - Ir$$

The effect of the internal resistance is to reduce the p.d. available to the external circuit. The p.d. across the internal resistance is often called the 'lost volts'.

As more current is drawn from the supply the 'lost volts' increase and the terminal p.d. falls.

If a high-resistance voltmeter is connected in parallel with a real cell, the current drawn is zero (an 'open circuit'), so the 'lost volts' is zero. The reading on the voltmeter is equal to the e.m.f. of the cell.

$$V_{\text{open circuit}} = \varepsilon$$

Real cells in circuits

Applying Kirchhoff's second law around the circuit:

$$\varepsilon = IR + Ir = I(R + r)$$

IR is the terminal p.d. applied to the external circuit, V.

$$\varepsilon = V + Ir \quad \text{or} \quad V = \varepsilon - Ir$$

If R is negligible (a short circuit), the terminal p.d. is zero. The short-circuit current is the maximum current that can be drawn from the cell. Car batteries, which have to supply in excess of 100 A, need a very low internal resistance.

Worked example

When a cell is connected across a $9.0\,\Omega$ resistor a current of $0.15\,A$ flows through the resistor. When the resistance is reduced to $4.0\,\Omega$ the current rises to $0.30\,A$.

Calculate the e.m.f. and internal resistance of the cell. **(5 marks)**

① $\varepsilon = IR + Ir = 0.15 \times 9.0 + 0.15r$

② $\varepsilon = 0.30 \times 4.0 + 0.30r$

From ① $r = \dfrac{(\varepsilon - 1.35)}{0.15}$

Solve these as simultaneous equations.

Substitute into ②.

$\varepsilon = 1.2 + 2(\varepsilon - 1.35) = 1.2 + 2\varepsilon - 2.70$

$\varepsilon = 1.50\,V$

Substitute back into ①.

$r = \dfrac{(1.50 - 1.35)}{0.15} = 1.0\,\Omega$

Now try this

When a cell is connected across a $1.0\,\Omega$ resistor a current of $8.0\,A$ flows through the resistor. When the resistance is increased to $5.5\,\Omega$ the current is $2.0\,A$.

(a) Calculate the e.m.f. and internal resistance of the supply. **(5 marks)**

(b) Calculate the terminal p.d. when the supply is connected to a load resistor of resistance $2.5\,\Omega$. **(3 marks)**

(c) Calculate the maximum current that can be drawn from this supply. **(2 marks)**

Experimental determination of internal resistance

E.m.f. and internal resistance can be determined by measuring the terminal p.d. and current drawn from a real cell.

Experimental determination of and r

Set up a circuit like the one below with a variable resistor.

Procedure:

1 Measure the current, *I* and terminal p.d., *V*, for different values of resistance, *R*, across a wide range.

Resistance (Ω)	V (V)	I (A)

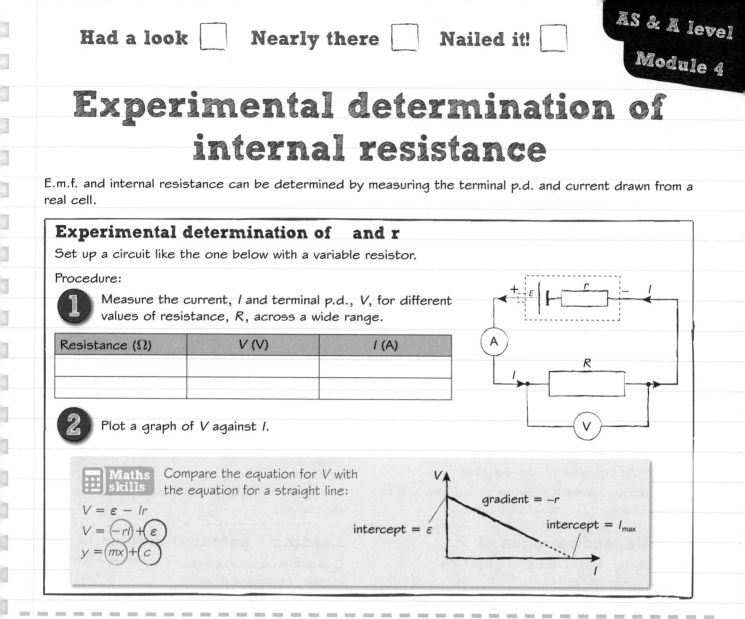

2 Plot a graph of *V* against *I*.

Maths skills Compare the equation for *V* with the equation for a straight line:

$V = \varepsilon - Ir$

$V = \boxed{-rI} + \boxed{\varepsilon}$

$y = \boxed{mx} + \boxed{c}$

intercept = ε

gradient = −r

intercept = I_{max}

Another graphical method

$V = \varepsilon - Ir$ can be rewritten in terms of *I* and *R*

$IR = \varepsilon - Ir$

This can be rearranged to give

$R = \dfrac{\varepsilon}{I} - r$

A graph of *R* against $\dfrac{1}{I}$ is also a straight line.

Its gradient is ε and the intercept on the *R* axis is −*r*.

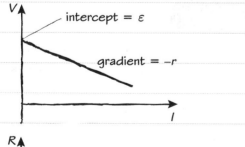

intercept = ε

gradient = −r

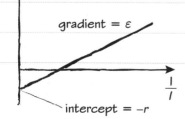

gradient = ε

intercept = −r

Now try this

Here are some data from an experiment like the one described above.

Resistance (Ω)	Terminal p.d. (V)	Current (A)
5.0	2.3	0.46
10.0	2.6	0.26
15.0	2.7	0.18
20.0	2.8	0.14
30.0	2.9	0.095

(a) Plot a graph of *V* against *I* and use it to determine the e.m.f. and internal resistance of the cell. **(6 marks)**

(b) Plot a graph of $\dfrac{1}{I}$ against *R* and use it to determine the e.m.f. and internal resistance of the cell. **(6 marks)**

Potential dividers

Potential divider circuits share the voltage from a supply between two resistors and are used to provide a calculated fraction of the supply voltage.

The potential divider circuit

This circuit is used to provide a voltage V_{out} from a supply voltage V_{in}.

The current through each resistor is the same.

$$I = \frac{V_{in}}{R_1 + R_2} \quad ①$$

$$I = \frac{V_{out}}{R_1} \quad ②$$

Equating ① and ②

$$\frac{V_{out}}{R_1} = \frac{V_{in}}{R_1 + R_2}$$

$$V_{out} = \frac{V_{in} \times R_1}{R_1 + R_2}$$

This is the potential divider equation.

The output voltage V_{out} can be varied from $0\,V$ to V_{in} by changing the resistance ratio.

Voltage ratio and resistance ratio

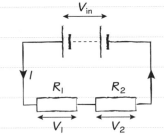

The current through each resistor is the same.

$$I = \frac{V_1}{R_1} = \frac{V_2}{R_2}$$

$$\frac{V_1}{R_1} = \frac{V_2}{R_2}$$

$$\frac{V_1}{V_2} = \frac{R_1}{R_2}$$

The voltage is split in the same ratio as the resistances.

Continuous variation of V_{out}

The resistor AB could be a long piece of resistance wire. The sliding contact means it can be considered as two resistors, one on each side of the contact.

As the sliding contact is moved from A to B, V_{out} changes continuously from $0\,V$ to V_{in}.

$$V_{out} = \left(\frac{x}{l}\right) V_{in}$$

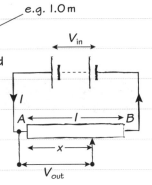

e.g. 1.0 m

Loading a potential divider

Connecting a load resistor — e.g. a lamp across the output of a potential divider changes the output voltage.

The left-hand side of the potential divider is now a lower resistance because it consists of R_1 and R_3 in parallel.

$$R_{Parallel} = \frac{R_1 R_3}{R_1 + R_3}$$

$$V_{out} = \frac{V_{in} \times R_P}{R_P + R_2}$$

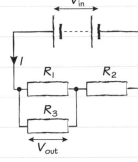

Worked example

For the circuit below, calculate the reading on a high-resistance voltmeter connected across (a) the $150\,\Omega$ resistor and then (b) the $450\,\Omega$ resistor. **(4 marks)**

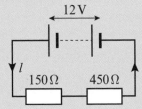

(a) When connected across the $150\,\Omega$ resistor

$$V_{out} = \frac{12 \times 150}{150 + 450} = 3\,V$$

(b) When connected across the $450\,\Omega$ resistor

$$V_{out} = 12 - 3 = 9\,V$$

Now try this

(a) Calculate the output voltage across the $250\,\Omega$ resistor in the circuit shown below. **(3 marks)**

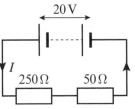

(b) A second $250\,\Omega$ resistor is connected in parallel with the first one in the potential divider above. Calculate the new output voltage across these two resistors. **(3 marks)**

Investigating potential divider circuits

Potential dividers are often used as the input to sensors.

Detecting light intensity

As light intensity increases, the resistance of a light-dependent resistor (LDR) decreases. If the LDR is used in a potential divider, the output p.d. will vary with light intensity.

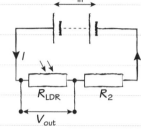

$$V_{out} = \frac{V_{in} \times R_{LDR}}{R_{LDR} + R_2}$$

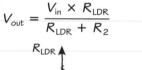

Light intensity　　　　Light intensity

If a voltmeter is connected across the output it can be calibrated to read in lux (light intensity units).

Designing a light sensor

In the circuit on the left, output p.d. decreases with increasing light intensity. However, light sensors are often used to switch on lighting when needed. A circuit in which p.d. increased with light intensity would be more useful.

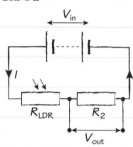

This can be achieved by using the p.d. output across R_2.

$$V_{out} = V_{in} - V_{LDR}$$

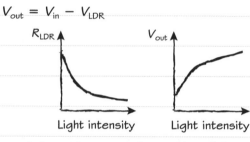

Light intensity　　　　Light intensity

The p.d. has a large gradient at low light intensities, so this sensor is more sensitive in that range and less sensitive to changes at high light levels.

A temperature sensor

The resistance of an NTC thermistor decreases as temperature rises, so it can be used in a potential divider as the basis of an electronic thermometer.

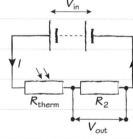

$$V_{out} = \frac{V_{in} \times R_2}{R_{therm} + R_2}$$

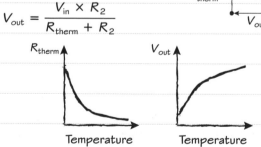

Temperature　　　　Temperature

The output can be connected to a high-resistance voltmeter and calibrated to use a temperature scale.

Worked example

A thermistor is used in a potential divider circuit like the one on the right, with a 6.0 V supply. It is in series with a fixed resistor of resistance 1 kΩ. The resistance of the thermistor varies from 5.0 kΩ at 10 °C to 100 Ω at 80 °C.

Calculate the p.d. across the fixed resistor (a) at 10 °C and (b) at 80 °C.　　**(6 marks)**

(a) At 10 °C the resistance of the thermistor is 5000 Ω.

$$V_{out} = \frac{V_{in} \times R_2}{R_{therm} + R_2}$$

$$V_{out} = \frac{6.0 \times 1000}{5000 + 1000} = 1.0 \text{ V}$$

(b) At 80 °C the resistance of the thermistor is 100 Ω.

$$V_{out} = \frac{V_{in} \times R_2}{R_{therm} + R_2}$$

$$V_{out} = \frac{6.0 \times 1000}{100 + 1000} = 5.5 \text{ V}$$

Now try this

An LDR is used as a light sensor in a potential divider circuit like the one on the top right of this page. Its resistance varies from 50 Ω in bright light to 2.5 kΩ in the dark. The supply p.d. is 5.0 V.

Calculate the range of p.d. outputs when the fixed resistor in series with the LDR is:

(a) 25 Ω　　**(3 marks)**　　　　(b) 500 Ω　　**(3 marks)**

(c) 10 kΩ.　　**(3 marks)**

Exam skills

This exam-style question uses knowledge and skills you have already revised. Have a look at pages 51–57 for a reminder about how to analyse DC electric circuits.

Worked example

The circuit below consists of a DC supply of e.m.f. 24 V, negligible internal resistance and three resistors.

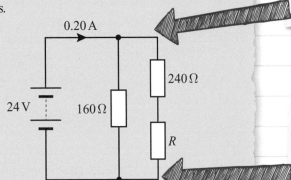

Two of the resistors have resistances 160 Ω and 240 Ω as shown.

The current drawn from the supply is 0.20 A.

(a) Calculate the resistance R. **(3 marks)**

The current through the 160 Ω resistor is
$I = \dfrac{24}{160} = 0.15\,A$.

The current through the other two resistors is therefore $0.20 - 0.15 = 0.05\,A$.

The total resistance of the two resistors on the right is $R_{tot} = \dfrac{24}{0.05} = 480\,\Omega$

Therefore $R = 480 - 240 = \underline{240\,\Omega}$

(b) Resistor R is now short-circuited by connecting a wire of negligible resistance in parallel with it.

State and explain what happens to the currents in each arm of the circuit when R is short-circuited. **(4 marks)**

There is no change in the current through the 160 Ω resistor. This is because it still has a potential difference of 24 V across it.
The current is $\dfrac{24}{160} = 0.15\,A$.

The current through the 240 Ω resistor increases. This is because the resistance of this arm has reduced but the potential difference across it has remained the same (24 V).

The current rises to $I = \dfrac{24}{240} = 0.10\,A$

Remember Kirchhoff's first law: 0.20 A enters this junction, so the sum of currents leaving it must also be 0.20 A. This will be useful when you work out the currents in the parallel arms.

The two branches of this circuit are in parallel with each other so they have the same potential difference across them (24 V). You can use this together with Ohm's law to calculate the current in each parallel arm.

Start by calculating the current through the 160 Ω resistor. This is connected directly across the supply, so it has a p.d. of 24 V across it.

This is where Kirchhoff's first law is used (see page 43 for a reminder of this important circuit law).

Notice how each step in the calculation has been presented in sequence and explained clearly – this helps to avoid making careless errors, and also shows that you fully understand what you are doing!

A short circuit has zero resistance. All current will flow through the short-circuit wire and not through R. The effect is the same as removing R and replacing it with the wire alone.

Command word: State

If a question asks you to 'state' something, just say what happens.

Command word: Explain

If a question asks you to 'explain' something, you should:

☑ write in full sentences

☑ explain each step in the process

☑ use correct scientific language.

Properties of progressive waves

What is a wave and how do we describe it?

Progressive waves (travelling waves)

Waves transfer energy from one place to another without any net transfer of matter.

Progressive waves move away from a source.

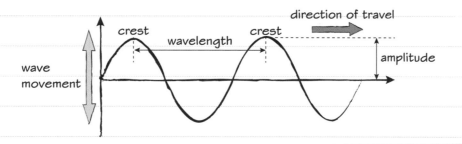

In a **transverse wave** the wave vibrations are perpendicular to the direction of energy transfer.

- Amplitude: the maximum displacement from equilibrium, that is, the distance from a peak or trough to the mean (rest) position; units m.
- Wavelength (λ): the distance between two successive identical points that have the same pattern of oscillation (vibrate in phase); units m.
- Wave speed (v): the speed at which a point of constant phase (e.g. a crest) moves; units m s^{-1}.
- Wave frequency (f): the number of oscillations per unit time at any point; units Hz.
- Time period (T): the time for one complete pattern of oscillation to take place at any point; units s.

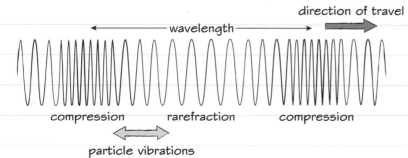

In a **longitudinal wave** the vibrations are parallel to the direction in which the wave travels.

Frequency and time period

Frequency is measured in hertz, a derived unit equivalent to s^{-1}.

  1 Hz = 1 wave per second.

Frequency and time period are related by the equation

$$f = \frac{1}{T}$$

Phase

As a wave passes a point, a particle at that point undergoes one complete cycle of vibration. It then repeats the motion. Since this motion is cyclic its position within the cycle is measured as an angle between 0° and 360° (or 0 radians and 2π radians).

Particles one wavelength apart oscillate in phase. Particles half a wavelength apart oscillate in antiphase (180 degrees or pi radians out of phase).

Now try this

Name two examples of transverse waves and two examples of longitudinal waves.   **(4 marks)**

The wave equation

Wave speed depends on frequency and wavelength.

Frequency, wavelength and time

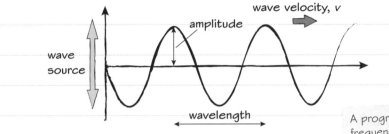

A progressive (travelling) wave of frequency f and wavelength λ

In time Δt there will be a total of $f\Delta t$ waves leaving the source.

The length Δx of the wave train will be $f\Delta t \times \lambda$.

The velocity of the wavefront will be $v = \dfrac{\Delta x}{\Delta t} = f\lambda$.

This is the **wave equation**

$$v = f\lambda$$

(units $\mathrm{m\,s^{-1}} = \mathrm{s^{-1}} \times \mathrm{m}$)

The speed of light

All electromagnetic waves travel at the speed of light, c, in a vacuum. This velocity is independent of the motion of the source or observer.

$$c = 299\,792\,458\,\mathrm{m\,s^{-1}} \approx 3.00 \times 10^8\,\mathrm{m\,s^{-1}}$$

The fixed velocity means that frequency and wavelength are inversely proportional to each other.

The speed of light is reduced in transparent materials. This results in refraction at a boundary. Longer wavelengths are refracted less. This results in dispersion (separation of white light into a spectrum of different colours).

The speed of sound

The velocity of sound in dry air at 20°C is $343\,\mathrm{m\,s^{-1}}$.

This velocity is measured relative to the air.

The speed of sound varies with temperature but not with the density or pressure of the air.

Ultrasound ($f > 20\,\mathrm{kHz}$) has the same speed in air as audible sound.

Worked example

(a) Humans can hear sounds in the range 20 Hz to 20 kHz. The speed of sound is $340\,\mathrm{m\,s^{-1}}$. Calculate the shortest wavelength of sound that can be heard by a human. **(2 marks)**

Shortest wavelength corresponds to highest frequency.

$$\lambda = \frac{v}{f} = \frac{340}{20\,000} = 0.017\,\mathrm{m} = 1.7\,\mathrm{cm}$$

(b) Visible light has a range of wavelengths of approximately 400–700 nm. Calculate the corresponding range of frequencies. **(4 marks)**

All light waves travel at $3.0 \times 10^8\,\mathrm{m\,s^{-1}}$

$$f = \frac{v}{\lambda}$$

At 400 nm $f = \dfrac{3.0 \times 10^8}{400 \times 10^{-9}} = 7.5 \times 10^{14}\,\mathrm{Hz}$

At 700 nm $f = \dfrac{3.0 \times 10^8}{700 \times 10^{-9}} = 4.3 \times 10^{14}\,\mathrm{Hz}$

Now try this

1　Calculate the frequency of an X-ray which has a wavelength of 10^{-10} m. **(2 marks)**

2　When light travels from air into glass its velocity reduces to $\frac{2}{3}$ of its value in air.
What happens to its frequency and wavelength? Explain your answer. **(4 marks)**

3　In an earthquake, P-waves (longitudinal) travelling through the Earth at $6.0\,\mathrm{km\,s^{-1}}$ and S-waves (transverse) travelling at $4.5\,\mathrm{km\,s^{-1}}$ leave the epicentre and are detected at a seismic station some distance away. The time between the arrival of the P- and S-waves is 15 s. How far away is the epicentre? **(4 marks)**

Graphical representation of waves

Wave displacement depends on more than one variable, so it is very important to think clearly about graphs of wave motion!

A transverse wave

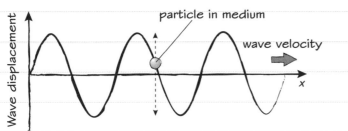

particle in medium

wave velocity

x

Wave displacement

A transverse wave moving through a medium – maybe a wave on a string. One particle has been magnified to show how it vibrates vertically up and down with the same frequency as the wave.

Two graphs can be drawn for the transverse wave.

Displacement

x

Displacement

t

1 Displacement vs x–position along wave

2 Displacement vs time for the particle

These graphs look the same, so make sure you read the axis labels carefully!

A longitudinal wave

Longitudinal waves vibrate along the direction in which the wave is travelling, but displacement can still be plotted on the y-axis of a graph. This graph looks the same as the graph for a transverse wave.

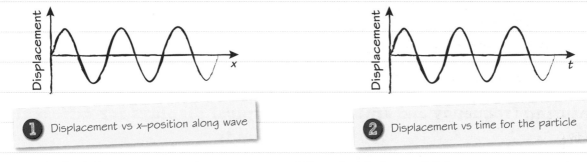

Displacement

x

λ

An alternative variable for the y-axis would be the variation in pressure above and below atmospheric pressure as the wave passes – the graph would look exactly the same.

As a sound wave passes through air, the air is compressed and rarefied in turn as particles move together and apart. The graph shows the particle displacement from its mean position.

Now try this

Two swimmers are floating 20 m apart in the sea as a water wave passes. The swimmers both complete one cycle of vertical oscillations in 5.0 s. The first swimmer notices that he is at the crest of a wave when the second swimmer is at the trough of the same wave.

(a) Calculate the frequency of the wave. **(2 marks)**

(b) State the wavelength of the water wave. Explain your answer. **(2 marks)**

(c) Calculate the wave speed. **(2 marks)**

Using an oscilloscope to display sound waves

Oscilloscopes display graphs of voltage against time.

Using an oscilloscope

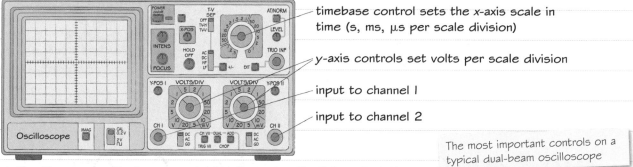

timebase control sets the x-axis scale in time (s, ms, μs per scale division)

y-axis controls set volts per scale division

input to channel 1

input to channel 2

Oscilloscope

The most important controls on a typical dual-beam oscilloscope

- Oscilloscopes display a graph of voltage (vertically) against time.
- Microphones transfer sound energy to electrical energy.
- If a microphone is connected to the input the graph represents vibrations in the sound wave.

Measuring frequency and amplitude

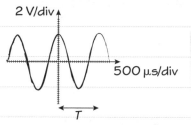

2 V/div

500 μs/div

T

The display produced when a microphone connected to an oscilloscope detects a sound.

The time period T of the oscillation is 4 divisions
= 4 × 500 μs = 2000 μs

Frequency f = $\frac{1}{T}$

$f = \dfrac{1}{(2000 \times 10^{-6})} = 500\,\text{Hz}$

The amplitude A of the oscillation is 2.5 divisions
= 2.5 × 2.0 = 5.0 V.

Typical displays for different sounds

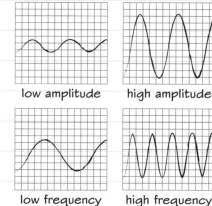

low amplitude   high amplitude

low frequency   high frequency

The timebase and voltage/division settings were the same for all of these displays.

Practical skills — Comparing waves and measuring phase difference

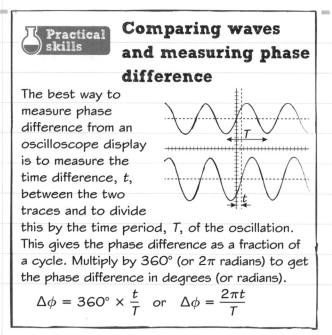

The best way to measure phase difference from an oscilloscope display is to measure the time difference, t, between the two traces and to divide this by the time period, T, of the oscillation. This gives the phase difference as a fraction of a cycle. Multiply by 360° (or 2π radians) to get the phase difference in degrees (or radians).

$$\Delta\phi = 360° \times \frac{t}{T} \quad \text{or} \quad \Delta\phi = \frac{2\pi t}{T}$$

Now try this

A sound is detected by a microphone attached to an oscilloscope.
The oscilloscope display shows a sinusoidal curve of amplitude 3.2 divisions,

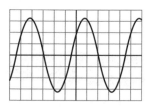

with one cycle of variation extending 5.0 divisions along the time axis. The oscilloscope settings are 50 mV/division vertically and 100 μs/division horizontally.

(a) Calculate the time period of the oscillation.
(2 marks)

(b) Calculate the frequency of the sound. **(2 marks)**

(c) What is the amplitude of the voltage variation? **(1 mark)**

Reflection, refraction and diffraction

These properties of waves can be seen clearly in a ripple tank.

The ripple tank

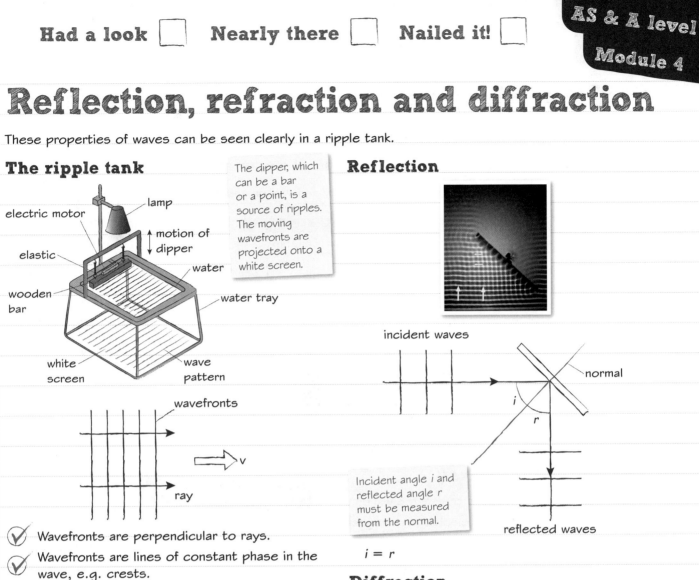

> The dipper, which can be a bar or a point, is a source of ripples. The moving wavefronts are projected onto a white screen.

electric motor
lamp
elastic
↕ motion of dipper
water
wooden bar
water tray
white screen
wave pattern

wavefronts

→ v

ray

✓ Wavefronts are perpendicular to rays.

✓ Wavefronts are lines of constant phase in the wave, e.g. crests.

Reflection

incident waves

normal

i

r

> Incident angle *i* and reflected angle *r* must be measured from the normal.

reflected waves

$i = r$

Diffraction

Diffraction is the spreading out of a wave after passing around an obstacle or through a gap.

> As waves pass through a gap (or past the edge of an object) they are diffracted into regions of geometric 'shadow'.

> The narrower the gap or the longer the wavelength, the greater the spread.

Diffraction is negligible if the gap or object is large compared with the wavelength.

Refraction

Refraction is a change in wave speed and sometimes direction that occurs when the wave moves from one material into another with different properties. If the wave direction is not perpendicular to the boundary, the wave direction will change as it crosses the boundary.

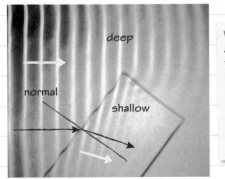

deep

normal

shallow

> Waves slow down as they cross the boundary between deeper and shallower water. This causes refraction: the slower ray refracts towards the normal.

Now try this

1 Which of the processes below explains why a swimming pool appears shallower than it actually is when you look down into it?　　　**(1 mark)**

　　A Reflection　　**B** Refraction　　**C** Diffraction　　**D** Polarisation

2 Explain why a loudspeaker can spread sound all around a room but a torch with the same diameter can only produce a beam of light.　　　**(4 marks)**

The electromagnetic spectrum

Visible light is a tiny part of the extended electromagnetic spectrum.

The electromagnetic spectrum – typical wavelengths and uses

increasing frequency and photon energy →

Radio	Microwave	Infrared	Visible light	Ultraviolet	X-ray	Gamma ray
$\lambda \approx$ m, km	$\lambda \approx$ mm, cm	$\lambda \approx \mu$m	$\lambda \approx 500$ nm	$\lambda \approx 100$ nm	$\lambda \approx 0.1$ nm	$\lambda \approx 10$ pm
telecommunications	mobile phones	remote controls	sight	disinfection	crystallography	radiotherapy

← increasing wavelength

ionising radiation

infrared light

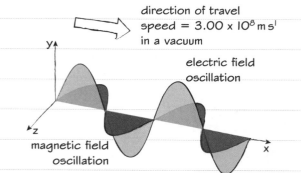

700 600   500   400
wavelength λ (nm)

Electromagnetic waves

• Electromagnetic waves are transverse waves.

• They are vibrations of the electric and magnetic field.

direction of travel
speed = 3.00×10^8 m s^{-1}
in a vacuum

electric field oscillation

y

z

magnetic field oscillation

x

• Electromagnetic waves are created by accelerating charges.

• When electromagnetic waves are absorbed by matter they make charges vibrate at the wave frequency.

Intensity

Intensity = power per unit area (units W m^{-2})

$$I = \frac{P}{A}$$

Intensity is directly proportional to wave amplitude squared.

$$I \propto (\text{amplitude})^2$$

Worked example

Exposure to ultraviolet light with a wavelength of less than about 310 nm burns skin. ⎯ called UVB

(a) Calculate the frequency of this light. **(2 marks)**

$$f = \frac{c}{\lambda} = \frac{3.00 \times 10^8}{310 \times 10^{-9}} = 9.7 \times 10^{14}\text{ Hz}$$

Sunlight in the ultraviolet A (UVA) range of 400–320 nm is thought to cause many skin cancers as well as tanning and wrinkling.

(b) Calculate the frequency range of UVA sunlight. **(3 marks)**

$$f = \frac{c}{\lambda} = \frac{3.00 \times 10^8}{320 \times 10^{-9}} = 9.4 \times 10^{14}\text{ Hz}$$

$$f = \frac{c}{\lambda} = \frac{3.00 \times 10^8}{400 \times 10^{-9}} = 7.5 \times 10^{14}\text{ Hz}$$

range = 7.5×10^{14} to 9.4×10^{14} Hz

(c) Is UVA more or less energetic than sunlight that causes sunburn? **(2 marks)**

UVA radiation has a longer wavelength (at least 320 nm) and therefore a lower photon energy than necessary to cause sunburn (about 310 nm).

See page 77 for more on photons.

Now try this

1 Calculate the frequency of red light of wavelength 700 nm. **(2 marks)**

2 Show that the intensity of light at a distance r from a point source that radiates equally in all directions with power P is inversely proportional to the distance squared, r^2. **(4 marks)**

Polarisation

Plane polarised waves are transverse waves that all vibrate in the same plane.

What is polarisation?

The particles in transverse waves vibrate at 90° to the wave's direction of travel. The vibration direction can lie anywhere on the plane perpendicular to the wave direction.

Unpolarised waves contain all possible vibration directions in the plane.

Plane polarised waves include just one direction (e.g. vertical or horizontal).

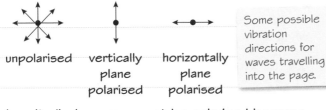

unpolarised · vertically plane polarised · horizontally plane polarised

Some possible vibration directions for waves travelling into the page.

Longitudinal waves cannot be polarised because they have a unique vibration direction.

parallel to their direction of travel

The fact that light and microwaves can be polarised shows that they must be transverse waves. Since electromagnetic waves consist of a magnetic field and an electric field oscillating at right angles to one another and to the direction of travel, we define the plane of electromagnetic waves as the plane of oscillation of the electric field.

Polarising filters

A polarising filter transmits light polarised in one direction and absorbs light polarised in the perpendicular direction.

Polarising filters for light are made of transparent polymers with long-chain molecules all aligned.

Light scattered from the sky is partially plane polarised. Light from the clouds is not. Cutting out some of the polarised light darkens the sky and increases contrast.

A metal grille acts as a polarising filter for microwaves. Here the vertical filter has absorbed the vertical component of the vibration and transmitted the horizontal component.

Practical skills ## Investigating polarisation of light

If two polarising filters are rotated with respect to each other, the amount of light transmitted varies as the angle between the two filters is changed.

The transmitted intensity is zero when the planes of polarisation of the two polarising filters are perpendicular and maximum when they are parallel to one another.

When unpolarised light falls onto a perfect plane polarising filter, the transmitted intensity is half the incident intensity.

The direction in which each filter transmits light is indicated by an arrow. As one filter is rotated, the amount of light reaching the meter changes.

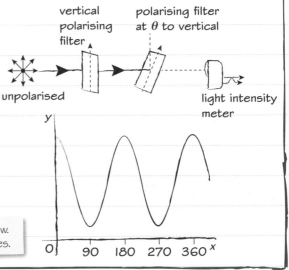

Now try this

Two polarising filters are crossed (their planes of polarisation are at 90°) and placed in front of a source of unpolarised light so that no light is transmitted. A third polarising filter is then slid between the original two so that its plane of polarisation is at 45° to the other two. Some light is now transmitted. Explain this effect.

(3 marks)

Refraction and total internal reflection of light

Refraction occurs when light enters a medium where its speed changes. This causes its direction to change (unless it enters normal to the boundary).

Refraction at a boundary

Refraction of light occurs because of a change of wave speed at a boundary between two transparent media (e.g. air and glass).

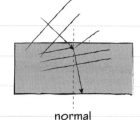

normal

The refractive index (n) of a transparent material is the ratio of the speed of light in a vacuum, c, to the speed of light in the material, v.

$$n = \frac{c}{v}$$

The greater the refractive index, the greater the change in speed (and hence direction) at the boundary.

Refraction of light as it leaves a transparent material

The refracted ray changes direction away from the normal as the light speeds up in air.

$$n = \frac{\sin\theta_2}{\sin\theta_1}$$

or

$$\frac{\sin\theta_1}{\sin\theta_2} = \frac{1}{n}$$

$\theta_2 > \theta_1$, so there is a maximum or critical angle θ_1 at which θ_2 reaches 90° and above which there is no refracted ray.

The law of refraction

$$n = \frac{\sin\theta_1}{\sin\theta_2}$$

The refractive index of all transparent materials is greater than 1, so $\theta_1 > \theta_2$ and the ray changes direction towards the normal when it travels from a vacuum into the material.

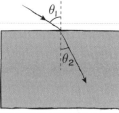

Total internal reflection

At and above the critical angle, $\theta_1 = C$, there is no refracted ray. All of the incident light is reflected inside the transparent material and total internal reflection occurs.

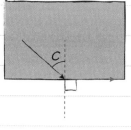

$$\frac{\sin\theta_1}{\sin\theta_2} = \frac{1}{n}$$

with $\theta_1 = C$ and $\theta_2 = 90°$

$$\frac{\sin C}{\sin 90°} = \frac{1}{n}$$

$$\sin C = \frac{1}{n}$$

Worked example

The diagram shows a ray of light approaching a rectangular block of glass of refractive index $n = 1.50$. Calculate the angles α, β, and γ, and comment on the direction of the emerging ray.   **(4 marks)**

$$n = \frac{\sin\theta_1}{\sin\theta_2}, \text{ so } 1.50 = \frac{\sin 60°}{\sin\alpha}$$

$$\sin\alpha = \frac{\sin 60°}{1.50} = 0.5774$$

$$\alpha = 35°$$

The two normals are parallel, so $\beta = \alpha = 35°$.

$$\frac{\sin\gamma}{\sin\beta} = 1.50$$

so $\sin\gamma = 1.50 \times \sin 35° = 0.8604$

$\gamma = 60°$ (after rounding)

The exiting ray is parallel to the original incident ray.

Now try this

1 A ray of light in air strikes a glass block at an angle of 43° to the normal to the surface. Calculate the angle of refraction inside the block if the refractive index of the glass is $n = 1.50$.   **(3 marks)**

2 A ray of light travels upward from the bottom of a pond and strikes the surface of water at 52° to the normal. The refractive index of water is 1.33. What happens to the light?
Explain your answer.   **(4 marks)**

3 Diamond has a critical angle of 25°. Calculate the refractive index of diamond.   **(2 marks)**

The principle of superposition

Superposition is a fundamental wave property that can explain interference and diffraction effects.

The principle of superposition

The resultant wave displacement at a point in space is equal to the vector sum of the displacements due to all of the waves that cross at that point.

All types of wave obey the principle of superposition.

Superposition in a ripple tank

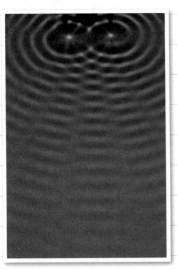

As water waves spread out from each dipper they superpose and create an interference pattern consisting of regions of **constructive interference** (reinforcement) and **destructive interference** (cancellation).

Simple examples

When two waves with the same frequency and amplitude superpose the resultant disturbance depends on their phase difference.

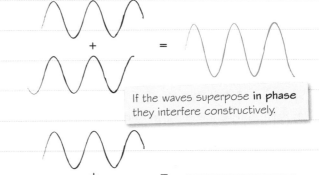

If the waves superpose **in phase** they interfere constructively.

If the waves superpose **in antiphase** they interfere destructively and cancel out.

Worked example

Sketch the resultant wave displacement of the superposition of waves A and B. **(3 marks)**

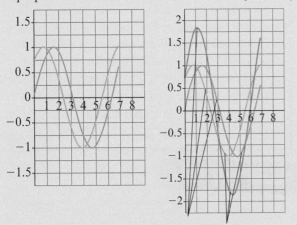

Add the values of the displacements for waves A and B to get the vaue for wave C at each point.

When the waves superpose the sign of the displacement is important!

The resultant is drawn by adding the displacements of each individual wave.

Now try this

The diagram below shows two waves travelling through the same region of space.

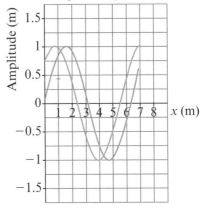

(a) Estimate the displacement of the resultant wave at $x = 0, 1, 2, 3, 4$ m **(5 marks)**

(b) Estimate the amplitude of the resultant wave. **(1 mark)**

(c) Copy the diagram above onto graph paper and sketch the resultant wave when these two waves superpose. **(5 marks)**

(d) Estimate the phase difference between the two individual waves. **(2 marks)**

Interference

Interference effects from wave superposition can be observed with every type of wave.

Interference with sound

Two loudspeakers are connected to the same audio frequency oscillator (AFO or signal generator).

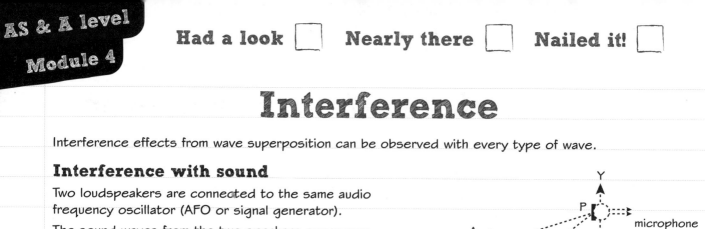

The sound waves from the two speakers superpose and form an interference pattern. As the microphone is moved from X to Y it moves through a sequence of maxima and minima of intensity.

The intensity detected at P depends on the **path difference** between the waves travelling from speakers A and B.

- Path difference = difference between the distances travelled by two waves arriving at the same point = BP − AP.

- If path difference is whole number of wavelengths – there is no **phase difference** between the waves (they interfere constructively and produce a maximum). — e.g. at the central position

- If path difference is odd number of half wavelengths – the interference is destructive, resulting in a minimum of intensity.

$$\frac{\lambda}{2}, \frac{3\lambda}{2}, \frac{5\lambda}{2}, \dots$$

Path difference BP − AP	Phase difference	Resultant
0	0	constructive interference
$\frac{\lambda}{4}$	$\frac{\pi}{2}$	intermediate intensity
$\frac{\lambda}{2}$	π	destructive interference
$\frac{3\lambda}{4}$	$\frac{3\pi}{2}$	intermediate intensity
1	2π	constructive interference

The shorter the wavelength, the closer together the maxima.

Interference between light waves

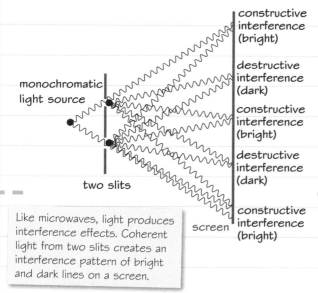

monochromatic light source

two slits

constructive interference (bright)

destructive interference (dark)

constructive interference (bright)

destructive interference (dark)

screen constructive interference (bright)

Like microwaves, light produces interference effects. Coherent light from two slits creates an interference pattern of bright and dark lines on a screen.

Monochromatic sources have a single wavelength.

For consistent interference the light must be **coherent** – the sources must maintain a constant phase (usually by being in phase).

Intensity at a maximum is 4× source

The intensity at a maximum in the pattern is 4 times the intensity at that point from just one of the sources. Interference means adding the amplitudes of the waves. Intensity depends on the square of the amplitude of the resultant wave.

$$\text{Amplitude}_{res} = \text{amplitude} + \text{amplitude}$$
$$= 2 \times \text{amplitude}$$

$$I_{res} = (2 \times \text{amplitude}_{res})^2 = 4 \times \text{amplitude}^2$$

Now try this

1 A teacher demonstrates interference effects using the set-up shown at the top of this page. As the microphone is moved further away from the central position the teacher notices that the minima have low intensity but are not zero. Suggest and explain a reason for this. **(3 marks)**

2 Suggest and explain two reasons why the illuminated region where light from two car headlamps superposes does not produce an interference pattern. **(4 marks)**

Two-source interference and the nature of light

A very useful equation can be derived for the separation of maxima and minima in an interference pattern created by equally spaced sources.

Analysing a two-source interference pattern

A and B are two coherent sources producing waves in phase at the source. A pattern of interference maxima and minima is formed along line XY.
- O is the position of the central maximum.
- P is the position of the first maximum from the centre.
- Path difference AP − BP = λ
- x is the separation of maxima.

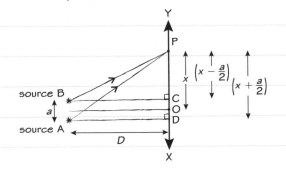

Using Pythagoras's theorem on triangles ADP and BCP, it can be shown that when the distance to the screen, D, is much larger than the source separation, a: ——— $D \gg a$

$$\lambda = \frac{ax}{D}$$

✓ For optical experiments, $D \gg a$ almost always holds (because the wavelength of light is so small).

✗ For experiments involving longer wavelengths (sound or microwaves), $D \gg a$ may not hold.

Measure the wavelength of light...

...by using $\lambda = \frac{ax}{D}$ and measuring the separation of the interference maxima (or **fringes**).

The nature of light

✓ Light is a wave – In the 17th century, Isaac Newton suggested that light consisted of tiny particles. Christiaan Huygens suggested that light consisted of waves. Superposition is a wave property. Diffraction and interference are evidence for the wave nature of light.

✓ Light is a wave – In 1801 Thomas Young passed coherent light through a double slit and measured the resulting interference pattern to calculate the wavelength of light.

✓ Light is a wave – Later in the 19th century, James Clerk Maxwell showed that light was an electromagnetic wave, part of a much wider spectrum including everything from radio waves to gamma rays.

✗ Light is not a wave! – 20th century discoveries about light's interaction with matter shows we sometimes have to use a particle model for light. ——— photons!

The current view is that light is described by quantum theory and is neither a wave nor a particle, although both models are useful for describing some of light's properties.

Light (like matter) exhibits **wave–particle duality**.

Worked example

Light of wavelength 500 nm is passed through a double slit apparatus with a slit separation a of 0.20 mm.

(a) Calculate the separation of maxima in the resultant interference pattern on a screen placed $D = 1.50$ m from the slits. **(2 marks)**

$$x = \frac{\lambda D}{a} = \frac{5.00 \times 10^{-7} \times 1.50}{2.0 \times 10^{-4}}$$
$$= 3.75 \times 10^{-3}\,\text{m} = 3.75\,\text{mm}.$$

(b) Describe the effect on the pattern of each of the changes below (applied separately).
  (i) Moving the screen back to a distance of 3.0 m. **(1 mark)**

Fringe separation will increase by a factor of 2 to 7.50 mm because $x \propto D$.

  (ii) Using light of wavelength 600 nm. **(1 mark)**

Fringe separation will increase by a factor of $\frac{6}{5}$ to 4.50 mm because $x \propto \lambda$.

  (iii) Replacing the double slit with another one of slit separation 0.40 mm. **(1 mark)**

Fringe separation will decrease by a factor of 2 to 1.88 mm because $x \propto \frac{1}{a}$.

Now try this

1 Light is passed through double slits of separation 0.42 mm. It forms an interference pattern on a screen at a distance of 2.40 m from the slits. The separation of the central maximum to the fourth maximum from the centre of the pattern is 1.19 cm. Calculate the wavelength of the light. **(4 marks)**

2 Explain why Young's double slit experiment provided evidence for the wave model of light. **(2 marks)**

Experimental determination of the wavelength of light

Thomas Young made the first measurement of the wavelength of light back in 1801; you can repeat this experiment and improve upon it in a school laboratory.

 Practical skills **Using a double slit**
To measure the wavelength of light use the equation $\lambda = \frac{ax}{D}$.

a = slit separation

D = distance between double slits and screen

x = separation of maxima (fringe separation)

1 Measure D with a metre rule.

2 Measure a with a travelling microscope or by using a projection method.

3 Find an average value for x by measuring across n fringes with a ruler and dividing by n.

The laser is a source of coherent monochromatic light.

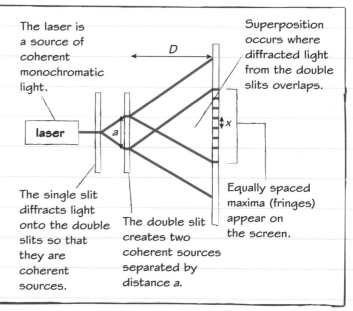

The single slit diffracts light onto the double slits so that they are coherent sources.

The double slit creates two coherent sources separated by distance a.

Superposition occurs where diffracted light from the double slits overlaps.

Equally spaced maxima (fringes) appear on the screen.

Different colours, different separations

Fringe separation $x \propto \lambda$, so with the same apparatus the fringes for red light would be further apart and the fringes for blue light would be closer together.

Measuring the fringes

12 fringe separations

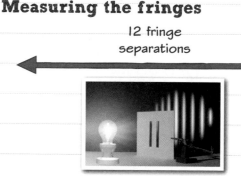

Worked example

A student sets up a double slit experiment using a red laser. She measures the distance to the screen as 0.96 ± 0.01 m and the slit separation as 0.18 ± 0.02 mm. She counts 15 fringes and finds the distance from the first fringe to the last to be 4.9 ± 0.1 cm.

Calculate the wavelength of the red light and state the uncertainty in your result. **(5 marks)**

$$x = \frac{4.9 \times 10^{-2}}{14} \text{ m} = 0.35 \times 10^{-2} \text{ m}$$

$$\lambda = \frac{ax}{D} = \frac{0.18 \times 10^{-3} \times 0.35 \times 10^{-2}}{0.96}$$

$$= 6.56 \times 10^{-7} \text{ m}$$

Now combine fractional uncertainties:

fractional uncertainty in λ

$$= \frac{0.01}{0.96} + \frac{0.02}{0.18} + \frac{0.1}{4.9} = 0.14 \ (14\%)$$

absolute uncertainty in λ

$$= 0.14 \times 6.56 \times 10^{-7} = 0.92 \times 10^{-7} \text{ m}$$

result: $\lambda = 6.56 \times 10^{-7}$ m ± 0.92×10^{-7} m

The fringes obtained with yellow light of wavelength 589 nm. 13 maxima are clearly visible, so the fringe separation is equal to the distance shown divided by 12.

Now try this

1 A student sets up a double slit experiment using a coherent source of yellow light. She measures the distance to the screen as 1.25 ± 0.01 m and the slit separation as 0.22 ± 0.02 mm. She counts 16 fringes and finds the distance from the first fringe to the last to be 5.00 ± 0.05 cm.

Calculate the wavelength of the yellow light and state the uncertainty in your result. **(5 marks)**

2 Explain why it is important in a double slit experiment:

(a) for the light source to be monochromatic

(b) for the two slits to produce coherent light. **(4 marks)**

Stationary waves

When two similar waves travelling in opposite directions superpose, a stationary wave pattern can form.

Formation of a stationary wave pattern

The conditions for the formation of a stationary (or 'standing') wave are:

1 the waves must be travelling in opposite directions

2 the waves must have equal or comparable amplitudes

3 the waves must be of the same type

4 the waves must have the same wavelength and frequency.

Under these conditions, superposition creates places where the waves always superpose in phase and interfere to create maxima, or **antinodes**, and places where the waves always superpose in antiphase and interfere destructively to create minima or **nodes**. The nodes and antinodes occur at fixed positions creating a **stationary** pattern.

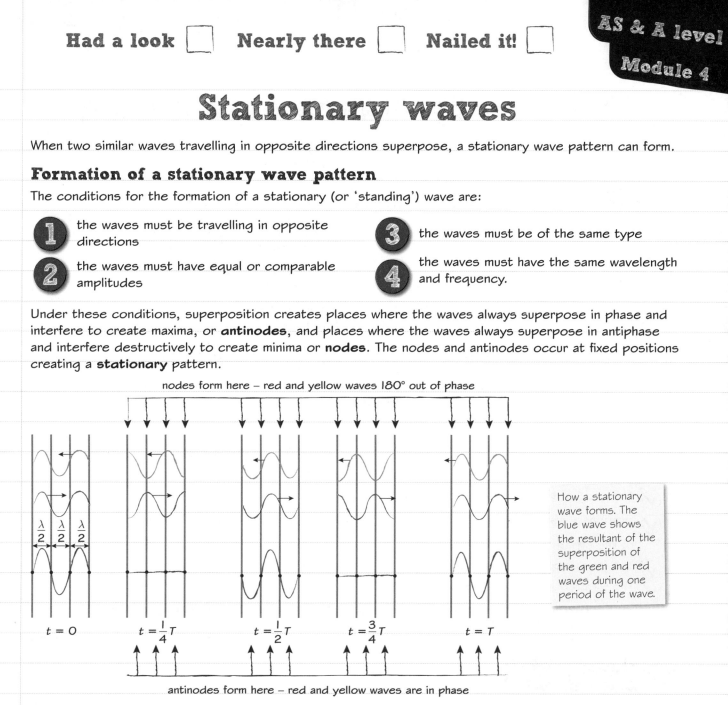

nodes form here – red and yellow waves 180° out of phase

$t = 0$ $t = \frac{1}{4}T$ $t = \frac{1}{2}T$ $t = \frac{3}{4}T$ $t = T$

antinodes form here – red and yellow waves are in phase

How a stationary wave forms. The blue wave shows the resultant of the superposition of the green and red waves during one period of the wave.

If both waves have amplitude A, then the amplitude at the antinodes will be 2A and the amplitude at the nodes will be zero.

The separation of adjacent nodes is $\frac{\lambda}{2}$.

Formation of a stationary wave by reflection

Reflected waves have the same wavelength and frequency as the incident wave and travel in the opposite direction, so they satisfy the conditions for the formation of a stationary wave. Most stationary waves form as a result of reflection.

Now try this

A stationary wave forms on a stretched wire as shown below.

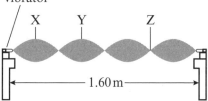

vibrator

X   Y   Z

←——— 1.60 m ———→

(a) State the wavelength of the wave and explain your answer. **(3 marks)**

(b) The frequency of the vibrations is 220 Hz. Calculate the speed of the waves on the string. **(2 marks)**

(c) Compare the amplitude and phase of vibrations at X and Y. **(2 marks)**

(d) Explain why the particles at Z do not vibrate. **(2 marks)**

(e) Explain why the stationary wave is formed on the wire. **(4 marks)**

Stationary waves on a string

The characteristic pitch and quality of sound from a stringed instrument depends on the stationary waves that are set up on the string when it is plucked, struck or bowed.

Practical skills: Investigating the stationary waves on a stretched string

mechanical vibrator of variable frequency

stretched string

fixed end

l

load to provide tension

fundamental (1st harmonic) — f

first overtone (2nd harmonic) — $2f$

second overtone (3rd harmonic) — $3f$

third overtone (4th harmonic) — $4f$

antinode   node

The vibrator is connected to a signal generator so that the frequency can be varied. Stationary waves appear at a discrete set of frequencies.

Stationary waves on a string. The lowest frequency at which a stationary wave is formed is called the **fundamental** or **first harmonic**. Higher harmonics occur at multiples of the fundamental frequency.

✓ The ends of the string are fixed points, so must always be nodes.

✓ Stationary waves can only be formed when an integer number of half wavelengths fits the length of the string: $\frac{n\lambda}{2} = 1, 2, 3, 4, \ldots$

✓ The frequency of each harmonic is given by $f = \frac{v}{\lambda}$.

✓ The speed of waves on the string is determined by the tension and mass per unit length of the string.

Increasing tension increases the speed and increasing the mass per unit length (for example by increasing the thickness) decreases the speed.

Worked example

The speed of transverse waves on a taut wire is $420\,\text{m s}^{-1}$. The length of the string is $62\,\text{cm}$. The string is plucked.

(a) Calculate the frequency of the fundamental vibration.   **(2 marks)**

For the fundamental vibration $\lambda = 2l$ so $\lambda = 124\,\text{cm} = 1.24\,\text{m}$.

$f = \frac{v}{\lambda} = \frac{420}{1.24} = 339\,\text{Hz}$

(b) Calculate the frequency of the third harmonic.   **(3 marks)**

For the third harmonic $\lambda = \frac{2}{3}l$ so $\lambda = \frac{2}{3} \times 0.62 = 0.413\,\text{m}$.

The speed of transverse waves is the same for all harmonics so $f = \frac{v}{\lambda} = \frac{420}{0.413} = 1017\,\text{Hz}$

Now try this

The A string of a guitar is $65\,\text{cm}$ long and has a fundamental frequency of $110\,\text{Hz}$. The speed of sound is $340\,\text{m s}^{-1}$.

(a) Calculate the speed of transverse waves on the string.   **(2 marks)**

(b) Calculate the frequency of the fourth harmonic on this string.   **(3 marks)**

(c) The guitar is left in a warm room. Explain why it goes out of tune and whether it becomes flat (lower) or sharp (higher).   **(4 marks)**

(d) Calculate the wavelength of the sound waves emitted from the string when it vibrates in its second harmonic mode.   **(3 marks)**

Had a look ☐ Nearly there ☐ Nailed it! ☐

Stationary sound waves

Like other waves, when sound waves are reflected and superposed they can form stationary waves.

Practical skills — **Using stationary waves to find the speed of sound**

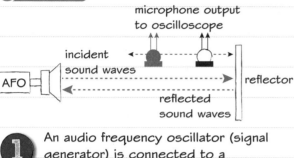

microphone output to oscilloscope

incident sound waves

AFO

reflector

reflected sound waves

1 An audio frequency oscillator (signal generator) is connected to a loudspeaker, which emits sound of frequency f.

2 A sound reflector is positioned such that the sound waves and their reflections combine to form a stationary wave.

3 A microphone is moved through the stationary wave pattern close to the reflector.

4 The oscilloscope displays how the amplitude of the sound varies with position.

5 The separation between adjacent antinodes (maximum amplitude positions) is $\frac{\lambda}{2}$.

6 Measure across several antinodes to reduce the error on the distance measured and calculate an average value of the wavelength.

7 Calculate the speed of sound from $v = f\lambda$.

A similar set-up can be used for microwaves. A suitable microwave reflector is a sheet of aluminium.

A resonance tube

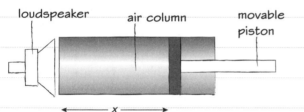

loudspeaker   air column   movable piston

x

The loudspeaker emits sound of frequency f.

Resonance occurs when the air column length, x, is such that a stationary wave forms in the tube. If the piston is pushed slowly into the tube there will be one or more positions at which a loud (resonant) sound is heard.

The distance, Δx, moved by the piston between one maximum and the next is half a wavelength.

the distance between two displacement nodes of the stationary wave

$\Delta x = \frac{\lambda}{2}$ — Again, λ can be used to calculate the speed of sound.

Worked example

A loudspeaker emits sound into one end of a resonance tube. A piston is slowly pushed into the other end of the tube. The sound intensity is a maximum when the piston has moved 3.0 cm, 23.0 cm and 43.0 cm from its starting position.

(a) Calculate the wavelength of the sound. **(2 marks)**

The distance between maxima is the distance between two adjacent nodes of the stationary wave.

This is equal to $\frac{\lambda}{2} = 20$ cm. So $\lambda = 40$ cm.

(b) Calculate the frequency of the sound, given that the speed of sound in the tube is 340 m s⁻¹. **(2 marks)**

$f = \frac{v}{\lambda} = \frac{340}{0.40} = 850$ Hz.

Now try this

The diagram below shows the pattern formed in sawdust placed in a resonance tube.

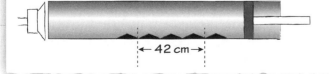

← 42 cm →

(a) Explain why the piles of sawdust form. **(3 marks)**

(b) Calculate the wavelength of the sound in the tube. **(3 marks)**

(c) Given that the speed of sound in the tube is 340 m s⁻¹, calculate the frequency of the sound in the tube. **(2 marks)**

Stationary waves in closed and open tubes

The stationary wave patterns in tubes depend on whether their ends are open or closed.

Boundary conditions

The conditions at the ends of a resonance tube are called boundary conditions.

 The end can be open.

 The end can be closed.

The boundary conditions determine whether the end of the tube is a node or antinode of particle displacement:

air column ←●→ open

particle antinode
at an open end

air column ●— closed

particle node
at a closed end

Stationary waves in tubes

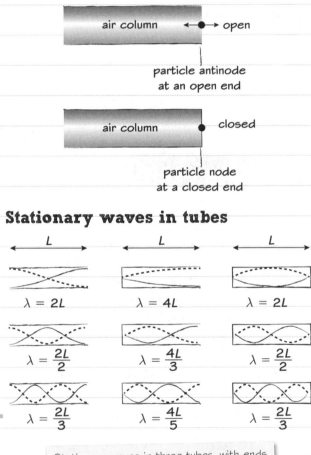

$\lambda = 2L$	$\lambda = 4L$	$\lambda = 2L$
$\lambda = \dfrac{2L}{2}$	$\lambda = \dfrac{4L}{3}$	$\lambda = \dfrac{2L}{2}$
$\lambda = \dfrac{2L}{3}$	$\lambda = \dfrac{4L}{5}$	$\lambda = \dfrac{2L}{3}$

Stationary waves in three tubes, with ends open/open, closed/open, closed/closed

- The sequence of wavelengths for the possible harmonics is the same for the tubes that have both ends open or both ends closed.

- The fundamental frequency of stationary waves in these tubes is $f = \dfrac{v}{2L}$.

- The sequence of frequencies is: f, $2f$, $3f$, ... like the string fixed at both ends.

- For the resonance tube open at one end the sequence is different: f, $3f$, $5f$. Only the odd harmonics of the fundamental frequency are produced.

Diagrams of stationary sound waves

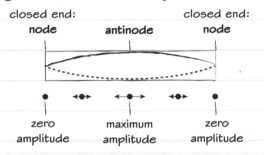

closed end: node   antinode   closed end: node

zero amplitude maximum amplitude zero amplitude

Amplitude of a longitudinal wave. Particle vibrations (directions and amplitudes) are shown beneath the tube.

Sound is a longitudinal wave, so the particle vibrations are parallel to the direction in which the wave travels. However, it is possible to represent a stationary wave using a diagram that shows how the vibration amplitude varies along the tube. It looks like a transverse wave but it represents a longitudinal wave!

Particle displacements and pressures

Displacement antinodes are pressure nodes, and displacement nodes are pressure antinodes.

time = t

time = $t + \dfrac{T}{2}$

displacement node

pressure antinode

The displacements of particles at two adjacent antinodes at two times, half a period apart

When the two particles approach one another the pressure increases. When they move away from one another the pressure drops.

Now try this

A student intends to produce stationary sound waves in a hollow tube of length 90 cm. The speed of sound in the tube is 340 m s⁻¹.

(a) Calculate the fundamental frequency in this tube when:

   (i)   the tube is open at both ends   **(2 marks)**

   (ii) the tube is open at one end and closed at the other end   **(2 marks)**

   (iii) the tube is closed at both ends.   **(2 marks)**

(b) Calculate the values of the first three lowest frequencies that resonate in the tube when it is closed at one end.   **(3 marks)**

Exam skills

This exam-style question uses knowledge and skills you have already revised. Have a look at pages 63 and 66 for a reminder about waves and their properties.

Worked example

The figure below shows a ray of light travelling from air into water. The refractive index of water is 1.33.

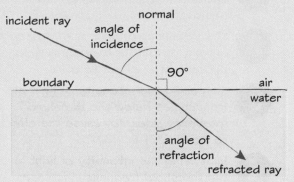

(a) Explain why light refracts at the boundary between air and water. **(2 marks)**

The speed of light in water is lower than the speed of light in air. Waves refract when they cross a boundary and the wave speed changes.

(b) The speed of light in air is $3.00 \times 10^8 \, \text{m s}^{-1}$. Calculate the speed of light in water. **(2 marks)**

Speed of light in water

$$c_{\text{water}} = \frac{c}{n} = \frac{3.0 \times 10^8}{1.33} = 2.26 \times 10^8 \, \text{m s}^{-1}$$

(c) A particular ray has an angle of incidence equal to 70°. Calculate the angle of refraction. **(3 marks)**

At the boundary

$$\frac{\sin i}{\sin r} = n$$

$$\sin r = \frac{\sin i}{n} = \frac{\sin 70°}{1.33} = 0.707$$

Therefore $r = 45.0°$

(d) Now consider rays of light travelling from water into air. Explain why there is a maximum angle of incidence (the critical angle) at the water/air boundary above which there will be no refracted ray. **(3 marks)**

Rays striking the water/air boundary refract AWAY from the normal. This means that there will be an incident angle less than 90° for which the refracted angle will be 90°. For incident angles larger than this critical value there can be no refracted ray; instead, total internal reflection occurs.

(e) Calculate the critical angle C at the water/air boundary. **(2 marks)**

$$\frac{\sin C}{\sin 90°} = \frac{1}{1.33} = 0.752$$

Critical angle $C = 49°$ (2 significant figures)

(f) Draw a diagram to show what happens when a ray of light travelling in the water strikes the water/air boundary at an incident angle greater than the critical angle. **(2 marks)**

Total internal reflection occurs as shown below:

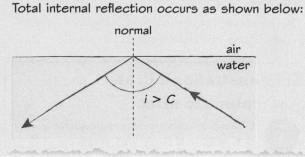

Diagrams

If you use a diagram to answer a question make sure that:

- ✓ it is relevant
- ✓ it is labelled
- ✓ it is drawn carefully (e.g. using a ruler and pencil).

Quote the equation you are using before you substitute values.

Make sure you start with a relevant equation and then show each stage of the working.

Here you might be tempted to say that total internal reflection will occur, but this doesn't answer the question. You are being asked why there is no refracted ray above a certain critical angle. Make sure your answer is focused on the question that is actually asked!

Don't confuse C, for critical angle, with c, for speed of light!

The photoelectric effect

Light falling on a metal surface can eject electrons from it, in a way that tells us a lot about the nature of light.

The photoelectric effect

The photoelectric effect is the ejection of electrons from a metal surface when light of a high enough frequency falls onto the surface.

Visible light does not discharge the electroscope, however intense it is.

If the electroscope is positively charged, even UV light does not discharge it.

Key observations of the demonstration

 Light can discharge a negatively charged electroscope, but not a positively charged electroscope.

 The frequency of light must exceed a certain **threshold frequency** f_0 for the photoelectric effect to occur.

3 Increasing the **intensity** of light at frequencies below the threshold frequency does not cause the effect to occur.

4 Increasing the **intensity** of light above the threshold frequency increases the rate of emission of electrons in direct proportion to the intensity but has no effect on their maximum kinetic energy.

5 Increasing the frequency of the light above the threshold frequency increases the maximum kinetic energy of the ejected electrons.

6 The value of the threshold frequency depends on the metal.

7 As long as $f > f_0$ the effect occurs as soon as light hits the metal surface.

Demonstration of the photoelectric effect

 Charge an electroscope negatively – the gold leaf rises and stays up.

2 Shine UV light at the zinc plate.

 The gold leaf falls – the electroscope is discharged.

UV light

zinc cap

gold leaf

When the metal cap and post of the electroscope are charged, the gold leaf is repelled and rises. It falls when UV light shines on the cap.

Problems for the wave model of light

We have found plenty of evidence to support the idea that light is a wave. The photoelectric effect cannot be explained by this model.

According to the wave model:

(X) Photoelectric emission should depend on intensity, not frequency.

(X) Photoelectric emission should not occur immediately – energy is spread evenly across the wavefront, so it should take time to build up enough to eject electrons.

(X) The energy of the emitted electrons should not depend on the frequency.

A photocell

Photocells transfer light energy to electrical energy in a circuit. They consist of a metal emitter and a collector. When light falls onto the emitter, electrons jump across to the collector and an e.m.f. is applied to the external circuit

photocurrent

Now try this

1 Give two pieces of evidence that support the wave model of light. **(2 marks)**

2 Explain why, if light is a wave, the photoelectric effect should depend on intensity and not frequency. **(2 marks)**

3 Suggest a reason why ultraviolet light can eject electrons from a zinc plate but red light cannot. **(2 marks)**

Einstein's photoelectric equation

Einstein realised that the photoelectric effect could be explained if we were prepared to consider a particle model for the nature of light.

Photons

Einstein proposed a particle model of light to explain the photoelectric effect.

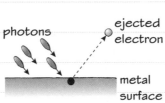

photons

ejected electron

metal surface

The key idea is that when light transfers energy to matter it can only do so in discrete amounts called **quanta**. These quanta of energy are called **photons**.

Photon energy $E = hf$

where h is the Planck constant, $h = 6.63 \times 10^{-34}\,\text{J s}$.

Using electronvolts

Work functions are often quoted in electronvolts. An electronvolt (eV) is a non-S.I. unit for energy. It is equal to the energy transferred when a charge e moves through a p.d. of $1.00\,\text{V}$.

$$E = QV = 1.60 \times 10^{-19}\,\text{C} \times 1.00\,\text{V}$$
$$= 1.60 \times 10^{-19}\,\text{J}$$
$$1\,\text{eV} = 1.60 \times 10^{-19}\,\text{J}$$

Electronvolts are often used to measure energies in atomic and particle physics.

Einstein's equation

$$hf = \phi + KE_{max}$$

This is the energy of the photon. All of this energy is transferred to ONE electron.

This is the work function of the metal.

This is the maximum kinetic energy of the ejected electron. It is equal to the difference between the photon energy and the work function.

Einstein assumed that one photon transfers all of its energy to one electron.

Changing light intensity changes the number of photons per second but does not affect their energy.

Explaining the photoelectric effect

Assume:

 There is a minimum amount of energy needed to eject an electron from a particular metal surface. This is called the **work function** and is given the symbol ϕ.

 The photons arrive randomly across the metal surface and each photon give all of its energy to a single electron in the surface.

Threshold frequency

If the photon energy is less than ϕ none of the photons can eject electrons. Their energy is dispersed by heating. Threshold frequency f_0 is given by: $hf_0 = \phi$

Electron kinetic energy

If the photon energy is greater than the work function, the electron is ejected with some kinetic energy: $KE_{max} = hf - \phi$

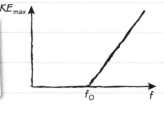

The maximum kinetic energy of photoelectrons is directly proportional to the photon frequency above the threshold frequency f_0.

Intensity

The more intense the light, the more photons reach the metal surface every second. This will increase the number of electrons leaving per second, but will not affect their kinetic energy.

Worked example

(a) Explain why the value of kinetic energy in Einstein's equation is a 'maximum' value. **(3 marks)**

The work function is the minimum energy needed to eject an electron from the surface of a metal. The actual energy required to eject a particular electron might be greater than this, for example if it is scattered by some of the other free electrons as it leaves the surface so that some of its energy is transferred away. In this case it will leave with a lower kinetic energy.

(b) Show that the maximum kinetic energy of photoelectrons emitted from a metal with threshold frequency f_0 is directly proportional to $(f - f_0)$. **(3 marks)**

$KE_{max} = hf - \phi$
But $\phi = hf_0$, so $KE_{max} = hf - hf_0 = h(f - f_0)$
Therefore $KE_{max} \propto (f - f_0)$.

Now try this

A metal has work function $\phi = 2.50\,\text{eV}$.

(a) Calculate the value of the work function in joules. **(1 mark)**

(b) Calculate the threshold frequency for this metal. **(2 marks)**

(c) Calculate the maximum kinetic energy of photoelectrons emitted from the surface of this metal by violet light of wavelength $420\,\text{nm}$. **(4 marks)**

Determining the Planck constant

The Planck constant is one of the fundamental constants of nature; its size determines the scale on which quantum effects are significant.

Using LEDs to determine the Planck constant

The voltage at which an LED lights up can be used to calculate a value for the Planck constant.

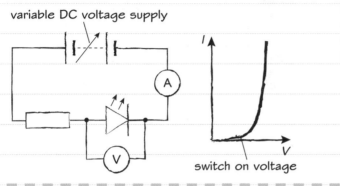

variable DC voltage supply

switch on voltage

When the forward voltage across a diode is increased it does not conduct until the voltage reaches a certain voltage, which depends on the material in the diode. Then the diode conducts with low resistance. This switch-on voltage can be detected in two ways.

1 Observing the diode through a long tube and noting the minimum voltage at which it can be seen to emit light.

2 Using a sensitive ammeter in the circuit and noting the voltage at which the current suddenly starts to rise.

Energy considerations

The voltage, V, across the LED measures the energy transfer per unit charge passing through the LED.

The energy transferred when one electron passes through the LED is given by

Energy transfer = charge × voltage = eV

When the LED just switches on, this energy is transferred to a single photon, so

$$eV = hf = \frac{hc}{\lambda}$$

We can calculate the Planck constant, h, if we use an LED that emits light of a known (or measured) wavelength and measure the switch-on voltage for the LED.

Measuring h

1 Set up a circuit like the one shown above.

2 Measure the switch-on voltage for a coloured LED.

3 Repeat for LEDs of different colours.

different wavelengths emitted

4 Record V and λ and plot a graph of V against $\frac{1}{\lambda}$.

5 $V = \left(\frac{hc}{e}\right) \times \frac{1}{\lambda}$

The graph should be a straight line through the origin with a gradient $\left(\frac{hc}{e}\right)$.

$V_{\text{switch-on}}$

$\frac{1}{\lambda}$

6 $h = \text{gradient} \times \left(\frac{e}{c}\right)$

Now try this

A student carried out an experiment to measure the Planck constant by measuring the switch-on voltage of several LEDs that emit light of different colours. The results are in the table below:

LED colour	Wavelength (nm)	Switch-on voltage (V)
red	635	1.96
orange	612	2.03
yellow	585	2.12
green	555	2.24
blue-green	505	2.46
ultra-blue	430	2.89

Plot a suitable graph and use it to determine the Planck constant.   **(6 marks)**

Electron diffraction

The discovery of electron diffraction showed that matter, like light, has quantum properties.

Demonstrating electron diffraction

The strongest experimental evidence for waves comes from superposition effects. These effects can be demonstrated for electrons by directing a beam of electrons at a thin slice of a polycrystalline material such as graphite.

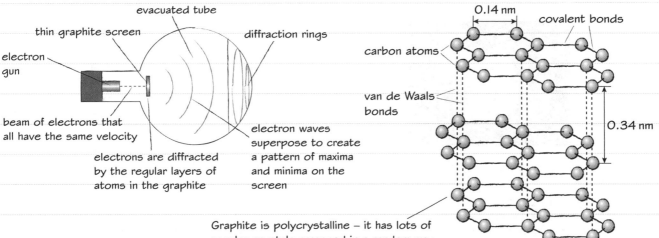

Graphite is polycrystalline – it has lots of regular crystals arranged in a random way.

Each graphite crystal has a structure of parallel layers of atoms that are the right distance apart to act like the slits in a diffraction grating. The electron waves are diffracted, and constructive and destructive interference occurs at particular angles.

Many such crystals are arranged in all orientations in each piece of graphite, so the maxima form rings of equal angle on the screen.

The de Broglie equation

De Broglie's equation links the particle property of electrons, momentum, to their wave property, wavelength.

$$\lambda = \frac{h}{p}$$

The faster the electrons are moving, the greater their momentum and the shorter their wavelength.

De Broglie's equation applies to all matter, but the effects are only really noticeable on the atomic scale.

Worked example

(a) Calculate the de Broglie wavelength of an electron travelling at 2% of the speed of light ($c = 3.0 \times 10^8\,\mathrm{m\,s^{-1}}$, $m_e = 9.1 \times 10^{-31}\,\mathrm{kg}$.) **(2 marks)**

$$\lambda = \frac{h}{mv} = \frac{6.63 \times 10^{-34}}{9.1 \times 10^{-31} \times 0.02 \times 3.0 \times 10^8}$$
$$= 1.2 \times 10^{-10}\,\mathrm{m}.$$

(b) Derive an expression for the de Broglie wavelength of an electron that has been accelerated through a p.d. of V volts. **(3 marks)**

$$KE = \tfrac{1}{2}mv^2 = eV$$
$$m^2v^2 = 2meV$$
$$mv = \sqrt{(2meV)}$$
$$\lambda = \frac{h}{\sqrt{(2meV)}}$$

(c) Calculate the de Broglie wavelength of an electron that has been accelerated through a p.d. of 500 V. **(2 marks)**

$$\lambda = \frac{h}{\sqrt{(2meV)}}$$
$$= \frac{6.63 \times 10^{-34}}{\sqrt{(2 \times 9.1 \times 10^{-31} \times 1.6 \times 10^{-19} \times 500)}}$$
$$= 5.5 \times 10^{-11}\,\mathrm{m}$$

A useful equation

If $p = mv$ ——— momentum

then $\lambda = \dfrac{h}{mv}$

Also $KE = \tfrac{1}{2}mv^2 = \dfrac{p^2}{2m}$

This is useful for finding momentum when you know the kinetic energy.

Now try this

(a) Calculate the de Broglie wavelength of an electron that has been accelerated through a p.d. of 45 V. **(2 marks)**

(b) Calculate the de Broglie wavelength of an electron that has a kinetic energy of 2.0 eV. **(3 marks)**

(c) Explain why a proton accelerated through a p.d. V has a shorter de Broglie wavelength than an electron accelerated through the same p.d. **(4 marks)**

Wave-particle duality

The wave and particle models are both useful, but neither provides a full description of matter or radiation.

Evidence for the wave model of light – superposition effects

Young's double slits experiment produces an interference pattern.

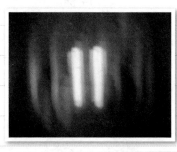

When waves superpose, the resultant intensity is the square of the sum of the wave amplitudes. If the waves arrive 180° out of phase they will interfere destructively and produce a dark band. If they arrive in phase they interfere constructively to produce a bright fringe.

Evidence for the particle model of light – the photoelectric effect

Einstein explained the photoelectric effect as a collision between a single photon and a single electron.

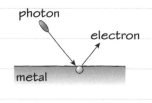

Energy is spread evenly over the wave front in a wave theory but in a particle theory the photons arrive randomly, with each one giving all of its energy to a single electron.

Wave theory predicts that electrons will be ejected if the intensity of the radiation is high enough. However, the effect only occurs above a threshold frequency. This can be explained if the light delivers its energy in quanta, or photons, with photon energy given by $E = hf$.

Evidence for the wave model of matter – superposition effects

When electron beams are fired at materials with a regular crystalline structure the scattering pattern has maxima and minima exactly as if the electron beam were short-wavelength waves.

Evidence for the particle model of the electron

- Ionic charges

When an atom gains or loses electrons its charge always changes by a multiple of the charge $e = 1.6 \times 10^{-19}$ C.

- Detection of beta particles

Beta particles are high-energy electrons. When they are detected by a Geiger–Müller tube they arrive as discrete entities – a fraction of an electron is never detected.

- Electrons also carry energy and momentum like a particle.

Wave-particle duality

Particle model – particles carry energy and momentum and are emitted and detected as discrete entities; we never detect a 'smeared out' electron or half a proton.

Wave model – waves have a measurable wavelength and obey the principle of superposition. This causes interference patterns to form when waves from two or more sources superpose.

The two models are linked by the de Broglie equation: $\lambda = \dfrac{h}{mv}$

Light and matter are both emitted and absorbed discretely, as particles.

When light or matter interacts with regular structure, diffraction occurs and a superposition pattern forms.

Both models are useful – but in different experimental circumstances.

Neither model is sufficient to explain everything about radiation or matter.

Now try this

Wave–particle duality is not restricted to the atomic world. In principle the de Broglie relation should apply to everything. Assuming that this is the case suggest and explain why wave properties such as diffraction do not present problems when trying to hit a tennis ball. **(6 marks)**

Exam skills

This exam-style question uses knowledge and skills you have already revised. Have a look at pages 76–80 for a reminder about the quantum nature of light.

Worked example

A red LED emits light of wavelength 630 nm. It is connected into a circuit like the one shown, and the supply voltage is slowly increased until the LED is just seen to glow. The reading on the voltmeter at this moment is 1.96 V.

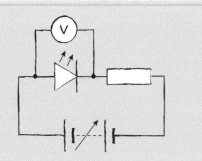

(a) Calculate the work done (in joules) on a single electron as it moves through the LED. **(2 marks)**

$W = eV = 1.60 \times 10^{-19} \times 1.96$
$\qquad = 3.14 \times 10^{-19}$ J.

(b) Assume that all of the energy supplied to the electron as it moves through the LED is transferred to one photon of red light of wavelength 630 nm. Use this assumption to calculate a value for the Planck constant.

(4 marks)

Photon energy $E = hf = \dfrac{hc}{\lambda}$.

If all of the work done on the electron is transferred to one photon, then
$E = 3.14 \times 10^{-19}$ J.

Rearranging to make h the subject:
$h = \dfrac{E\lambda}{c} = \dfrac{3.14 \times 10^{-19} \times 630 \times 10^{-9}}{3.0 \times 10^{8}}$
$\qquad = 6.59 \times 10^{-34}$ J s

(c) When the supply voltage is increased, the LED glows more brightly but the colour of the light it emits is unchanged. Explain this in terms of photons. **(4 marks)**

Increasing the supply voltage will increase the current flowing through the LED.
More electrons per second will pass through the LED, so more photons per second will be emitted. This accounts for the increase in brightness. The fact that the colour of the light does not change means that the wavelength is still the same, so the photons have the same energy as before. Therefore each electron is transferring the same amount of energy to a photon as it passes through the LED.

(d) Suggest what the observations in part (c) imply about how the potential differences across the LED and resistor change as the supply voltage is increased. **(3 marks)**

If the energy transferred by each electron is unchanged, then the voltage across the LED is unchanged. The sum of the p.d. across the LED and the p.d. across the resistor must equal the supply voltage, so the p.d. across the resistor must increase as the supply voltage increases.

Command word: Calculate

When you are asked to calculate something you should always show your working.

It is good practice to start with the algebra (rearranging equations if necessary), then substitute values and finally calculate your answer.

Don't forget to include the units! If you get stuck, you can work out the units from the equation.

Command word: Explain

If a question asks you to 'explain' something, make sure you:

✓ Address all parts of the question: in part (c) there are **two** things to explain.

✓ Respond to any directions in the question: in part (c) you are asked to explain **in terms of photons**, so make sure that is what you do.

✓ Think first and write your answer in a clear logical sequence.

Notice how the answer to part (d) has used Kirchhoff's second law (see page 51 for an explanation of this very useful circuit law).

Temperature and thermal equilibrium

If two or more bodies are in thermal equilibrium, then they are at the same temperature.

Kinetic theory of matter

The kinetic theory models all matter as molecules (or atoms or ions) in a continuous state of motion.

The molecules have kinetic energy due to their random motion, and potential energy due to forces between the atoms or molecules. These energies are randomly distributed between molecules.

Internal energy is the sum of the kinetic and potential energies of the molecules. It depends on the temperature, but internal energy and temperature are not directly related unless the number of molecules is constant.

🧪 Practical skills Brownian motion

Evidence for the continual random motion of molecules in liquids and gases comes from Brownian motion.

microscope
smoke particles
air cell

Smoke particles in the air cell are buffeted by air molecules. Under the microscope the smoke particles can be seen as erratically moving dots of light.

Temperature

Temperature is the degree of hotness of a body. It is measured by a thermometer.

Thermometers have a property that varies with temperature (a thermometric property) such as a change in liquid volume or a change in resistivity.

The thermometric property and the scale chosen will determine the numerical value of the temperature for a given situation.

Thermal equilibrium

Two bodies in thermal equilibrium: no net transfer of energy between them, bodies have the same temperature.

Two bodies at different temperatures: energy transfer from the higher to the lower temperature body.

Temperature scales

To set up a temperature scale the thermometer must be calibrated.

On the Celsius scale the calibration points are the melting point and the boiling point of water.

The thermometer is allowed to reach thermal equilibrium with another body at the calibration points, and its thermometric property measured each time. As both bodies are at the same temperature, the new thermometric property value indicates the temperature.

°C
120 110 100 90 80 70 60 50 40 30 20 10 0 -10

The Kelvin temperature scale

The thermodynamic (Kelvin) scale is an absolute scale of temperature, as it does not depend on the property of any particular substance.

Absolute zero is the temperature at which the internal energy is a minimum.

Divisions on the Kelvin scale are the same size as divisions on the Celsius scale. To convert from Celsius temperature, θ, to an absolute temperature, T, requires a simple addition.

$T = \theta + 273$

$0\,°C = 273\,K$

⬅ The symbol for the unit kelvin is K and not °K.

Worked example

The pressure p exerted by a fixed mass of air maintained at a constant volume is directly proportional to the absolute temperature T of the gas. At a temperature of 273 K the pressure exerted by the air is 98 kPa.

Calculate the temperature when the pressure exerted by the air is 120 kPa. **(2 marks)**

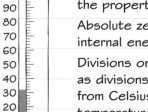

$p \propto T$

$\therefore \dfrac{p_1}{p_2} = \dfrac{T_1}{T_2} \quad \therefore T_2 = 273 \times \dfrac{120}{98} = 334\,K$

Now try this

Describe an experiment that demonstrates Brownian motion and discuss how this provides evidence for the movement of molecules in a fluid. **(6 marks)**

Solids, liquids and gases

Internal energy depends upon temperature, the number of molecules and the way in which the molecules are ordered.

Intermolecular potential energy

Intermolecular potential energy arises because of the forces between molecules.

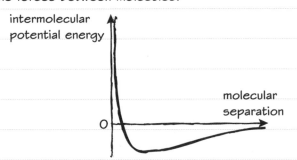

The intermolecular potential energy for a solid has a large negative value, as the molecules are close together. Energy has to be transferred into the solid to move the molecules apart.

The potential energy is less negative for liquids, as the molecules are further apart.

Potential energy is zero for an ideal gas, as the molecules are far apart.

Molecular separation

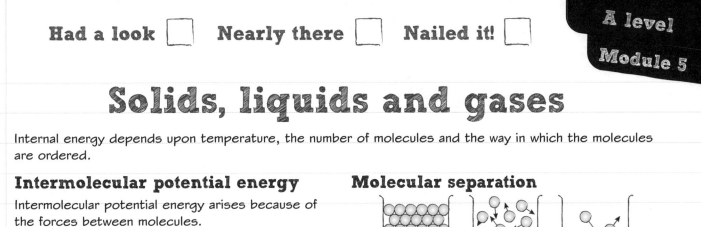

solid   liquid   gas

temperature

The molecules in a solid are in a highly ordered state. This gives solids a fixed volume and shape. At temperatures above absolute zero the molecules vibrate around relatively fixed positions. In a liquid, molecules are still close together, but in a less ordered state. The molecules are not constrained to fixed positions. Liquids thus have a fixed volume but can take up the shape of their containers.

In a gas, molecules are in a very disordered state. The molecules move independently of each other and the gas fills up any container it is released into.

Molecular kinetic energy

As the temperature is increased, the kinetic energy of the molecules increases. Hence the internal energy (the sum of the kinetic and potential energy) increases.

As a substance changes state the kinetic energy stays constant, but the potential energy increases (becomes less negative).

Changes of state

A beaker of ice from the freezer is heated at a constant rate and its temperature monitored.

As the ice melts the temperature of the ice/water mixture stays constant, as long as the mixture remains in thermal equilibrium. (For example, rapid, uneven heating faster than heat can be transferred through the mixture might raise the temperature in some areas.)

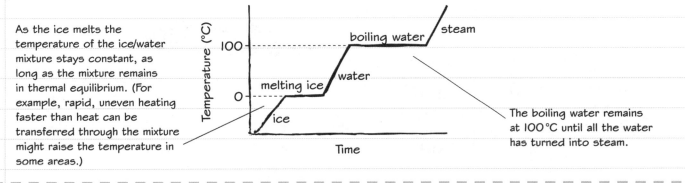

The boiling water remains at 100 °C until all the water has turned into steam.

Now try this

When a sample of ice melts to form water the temperature remains constant until all of the ice has melted. Explain on a molecular level what happens to the energy transferred to the sample during this melting process. **(3 marks)**

Specific heat capacity

When a sample is heated and the temperature rises, the temperature rise depends on what it is made of.

Heating and temperature

When energy is transferred to a sample by heating, its temperature will increase, unless it changes state.

Transferring energy at the same rate to different samples will not produce the same rate of temperature increase. The energy transfer results in an increase in the kinetic energy of the molecules (or atoms or ions) in the sample. So the number of molecules in the sample is a factor in the temperature increase. A large mass needs more energy transferred than a small one for the same temperature change.

Specific heat capacity

Transferring energy at the same rate to equal masses of different materials produces different rates of change of temperature.

We can define a **specific heat capacity**, *c*, to enable us to predict temperature changes for different materials. The specific heat capacity (s.h.c.) is the energy needed to increase the temperature of a 1 kg sample of material by 1 K (or 1 °C).

$$c = \frac{E}{m\Delta\theta} \text{ (units } J\,kg^{-1}\,K^{-1})$$

where E is the energy transferred to the sample of mass m, and $\Delta\theta$ is the temperature increase.

🧪 Practical skills — Determining the s.h.c. of a metal

A metal block has holes drilled into it so that a thermometer and immersion heater fit inside.

The block is surrounded by a layer of insulating material so that when energy is transferred to the block from the immersion heater little energy is transferred out to the surroundings.

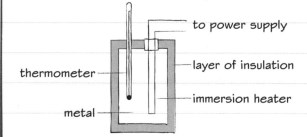

to power supply

layer of insulation

thermometer

immersion heater

metal

The current in the circuit and the p.d. across the heater must be measured so that the rate of energy transfer can be calculated.

🧪 Practical skills — Determining the s.h.c. of a liquid

A small quantity of liquid is heated in an insulated container with an electrical heater.

The heater is turned on for a known time and the temperature rise is measured.

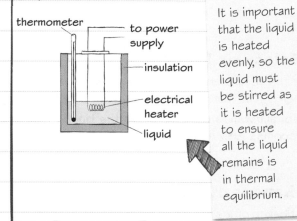

thermometer

to power supply

insulation

electrical heater

liquid

It is important that the liquid is heated evenly, so the liquid must be stirred as it is heated to ensure all the liquid remains is in thermal equilibrium.

Assume that the energy transferred from the heater and the block to the surroundings, is negligible. Always refer to the energy transfer process, and avoid making statements such as 'no energy is lost'. In fact both the surroundings will be heated too, so not all the energy transferred from the heater is raising the temperature of the aluminium.

Worked example

The temperature of a 990 g aluminium block increases by 10.5 °C when a 20 W heater is used to heat it. Calculate for how long the heater was switched on. ($c_{aluminium} = 912\,J\,kg^{-1}\,K^{-1}$.)   **(3 marks)**

$E = mc\Delta\theta \quad \therefore \quad Pt = mc\Delta\theta$

$t = \dfrac{0.99 \times 912 \times 10.5}{20} = 474\,s$

Now try this

A coffee maker heats 245 g of water from 12.5 °C to 95.0 °C to make a cup of coffee. Calculate the energy needed, stating the assumption that you are making.   **(3 marks)**

Specific heat capacity of water = 4180 J kg⁻¹ K⁻¹

Specific latent heat 1

When a sample is heated at its melting or boiling point its temperature stays constant.

Latent heat

Energy transfer to a sample at its melting or boiling point does not change the temperature. The average kinetic energy of the molecules remains constant during the change of state. Instead, the energy transfer causes an increase in the potential energy of the molecules in the sample. The energy transferred to the sample is referred to as latent (hidden) heat, as there is no change in temperature.

Melting or freezing, boiling or condensing?

When energy is transferred to a solid at its melting point, it melts. The energy is used to increase the potential energy of the molecules. In a similar way, when a liquid cools to its melting point it will solidify. Energy is transferred from the sample to the surroundings, and the potential energy of the molecules decreases.

We call the energy exchanged when 1 kg of a material melts or solidifies the **specific latent heat of fusion**, L_f.

The energy exchanged when 1 kg of a material boils or condenses is the **specific latent heat of vaporisation**, L_v.

For water:
$L_f = 334$ kJ kg^{-1} $L_v = 2\,260$ kJ kg^{-1}

Because the change in the potential energy of the molecules is much larger when the change is from liquid to gas (or vice versa), the latent heat of vaporisation is always considerably larger than the latent heat of fusion for a given material.

Specific latent heat

The energy required to change the state of 1 kg of the sample (solid to liquid, or liquid to gas) is referred to as the specific latent heat of the material, L.

$$L = \frac{E}{m} \text{ (units J kg}^{-1}\text{)}$$

where E is the energy transferred to the sample and m is the mass of sample that changes state.

Practical skills — Determining the specific latent heat of fusion of water

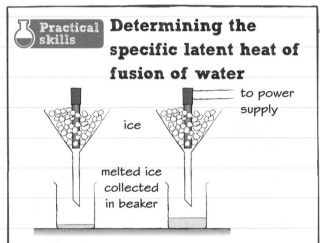

ice

to power supply

melted ice collected in beaker

Place the same mass of crushed ice in each funnel. Switch on the heater and start a timer. Record the heater p.d. V and current I.

Leave the heater on until some ice has melted. Switch off the heater and record the time, t. Measure the mass of water in each beaker.

Water collected from the funnel in which the heater was not switched on is ice melted by energy transfer from the surroundings.

$VIt = mL$, where m is the difference in the mass of melted ice for each beaker.

Worked example

Calculate the time taken for an 85 W heater to melt 1.5 kg of ice at 0 °C.   **(2 marks)**

$L_f = 330$ kJ kg^{-1}

$Pt = mL_f$ $\therefore$ $t = \dfrac{1.5 \times 3.3 \times 10^5}{85} = 5800$ s

Remember that when writing down equations you need to define all the symbols that you use. The equation $VIt = mL$ only makes sense if we have specified what each symbol represents.

This includes units: time t in seconds, for example.

Now try this

An experiment to determine the specific latent heat of fusion of water uses the technique illustrated above. A 1 kW heater takes 4.3 minutes to melt 800 g of ice at 0 °C.

(a) Calculate the specific latent heat of fusion.   **(2 marks)**

(b) Why is crushed ice used?   **(1 mark)**

(c) What happens to the value of L_f obtained if the mass of ice melted by energy transfer from the surroundings is not accounted for?   **(1 mark)**

(d) Suggest two reasons why the experimental result found here differs from the accurate result of 330 kJ kg^{-1}.   **(2 marks)**

Specific latent heat 2

There are many more methods for measuring specific latent heat.

🧪 Practical skills Determining the specific latent heat of fusion of water by the method of mixtures

Polystyrene cups with lids are used to minimise energy transfer to the surroundings. Sufficient hot water of known temperature is added to a known mass of ice at 0 °C to melt all of the ice. The final mass and temperature of the mixture is recorded.

The energy transferred from the hot water to the ice equals the energy needed to melt the ice plus the energy required to raise the temperature of the melted ice to the final temperature.

$$E = m_{water}c(\theta_1 - \theta_2) = m_{ice}L_f + m_{ice}c\theta_2$$

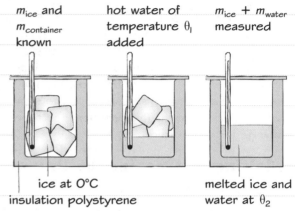

m_{ice} and $m_{container}$ known

hot water of temperature θ_1 added

$m_{ice} + m_{water}$ measured

ice at 0°C
insulation polystyrene

melted ice and water at θ_2

🧪 Practical skills Determining the specific latent heat of vaporisation of water

An immersion heater is used to boil water in an open polystyrene cup. The decrease in mass over a known time period is recorded. The current in the circuit and the p.d. across the immersion heater must be measured.

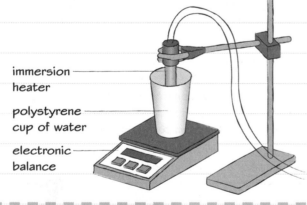

immersion heater

polystyrene cup of water

electronic balance

Calculating the specific latent heat of vaporisation of water

The energy transferred from the heater boils the water.

The electronic balance records the mass of the water and the polystyrene cup. Once the water is boiling at 100 °C readings can be taken. The mass registered on the balance is recorded and a stopwatch started.

Once the immersion heater has boiled off a sufficient mass of water, the heater is turned off and the stopwatch stopped.

The mass of water boiled off m is calculated from the difference in the two balance readings.

$$VIt = mL_v$$

Energy will be transferred to the surroundings, so the value obtained for L_v will be larger than the true value. The insulating polystyrene of the cup is one way of minimising this error. Another is using a high-power heater to decrease the time taken for a given mass of water to be boiled away. The shorter the time, the less energy transfer to the surroundings.

Worked example

Calculate the amount of energy transferred to the surroundings when 350 g of steam at 100 °C condenses and then cools to 40 °C.   **(3 marks)**

$L_v = 2.26\,\text{MJ kg}^{-1}$; $c_{water} = 4.18\,\text{kJ kg}^{-1}\,\text{K}^{-1}$

$E = mL_f + mc\Delta\theta$
  $= m(L_f + c\Delta\theta)$
  $= 0.35 \times (2.26 \times 10^6 + 4.18 \times 10^3 \times 60)$
  $= 878\,780\,\text{J} = 879\,\text{kJ}$

Be careful with units!

In this example mass must be converted to kg, L_v must be in J kg^{-1}, and c in $\text{J kg}^{-1}\text{K}^{-1}$.

$1\,\text{MJ} = 1 \times 10^6\,\text{J}$
$1\,\text{kJ} = 1 \times 10^3\,\text{J}$
$1\,\text{g} = 1 \times 10^{-3}\,\text{kg}$

Now try this

Explain why a scald by steam at 100 °C is much more painful than a scald by water at 100 °C.   **(3 marks)**

Kinetic theory of gases

In the kinetic theory, ideal gases are assumed to consist of molecules that are in continuous, random motion.

Kinetic theory

The kinetic theory starts from a number of basic assumptions about gases, which lead to the concept of an **ideal gas**. Then statistics and the laws of Newtonian mechanics are applied to the motion of the molecules.

As a result the large-scale properties of gases can be described in terms of the behaviour of molecules.

Assumptions of the kinetic theory

Gases consist of a large number of molecules or atoms in continuous, rapid, random motion.

The molecules occupy negligible volume compared with the volume of the gas.

Collisions between molecules and between the molecules and the walls of the container are perfectly elastic.

The time of the collisions is negligible compared with the time between collisions.

Intermolecular forces are negligible except during collisions between molecules.

Pressure

On a macroscopic scale, pressure p is defined as the force F per unit area A, $p = F/A$. The force is perpendicular to the area.

See page 20 for more about pressure.

On a microscopic scale, pressure arises as a result of molecular bombardment of the walls of the container.

The exchange of momentum that takes place between molecules and the container walls produces a force according to Newton's second law.

See page 35 for a reminder of Newton's laws of motion.

Temperature

On a macroscopic scale, temperature is the degree of hotness of a body.

On a microscopic scale, temperature is related to the average kinetic energy of the molecules.

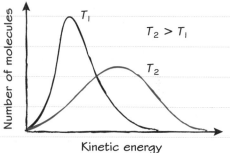

$T \propto \langle E_K \rangle$

where temperature T is measured on the absolute or Kelvin scale.

Worked example

Explain in terms of kinetic theory and molecular behaviour why a metal can collapses when air is pumped out of it with a vacuum pump. **(6 marks)**

As air is pumped out of the can the number of molecules of air decreases. Since the temperature doesn't change, the average kinetic energy of the air molecules stays constant. But the rate of collision of air molecules with the walls of the container decreases.

So the total rate of change of momentum as the molecules strike the walls of the container decreases. From Newton's second law, the force on the walls decreases, and hence the pressure inside the can decreases.

If atmospheric pressure outside the can exceeds the pressure inside the can, then forces will act to collapse the can.

Worked example

Calculate the change in momentum when a molecule of mass 2.6×10^{-26} kg travelling at $380\,\mathrm{m\,s^{-1}}$ makes a head-on elastic impact with a surface. **(2 marks)**

$m(v - u) = 2.6 \times 10^{-26} \times (380 - (-380))$
∴ momentum change = 1.98×10^{-23} Ns

Now try this

Calculate the net force exerted on a window of area $2.5\,\mathrm{m^2}$ when there is a pressure difference of 1500 Pa between the two sides of the window. **(2 marks)**

The gas laws: Boyle's law

The pressure exerted by a gas can be related to the volume the gas occupies.

Real gases

The gas laws are experimental laws that were found to be approximately true for **real gases** under certain conditions.

An **ideal gas** obeys the laws exactly under all conditions.

Ideal gases

Ideal gases are as specified by the assumptions of the kinetic theory (see page 87).

Real gases deviate from ideal behaviour under extremes of temperature and pressure because those assumptions are not perfectly true.

🔢 Maths skills — Boyle's law

Robert Boyle investigated how the volume occupied by a fixed mass of dry air depended upon the pressure. In his experiment the temperature of the gas was kept constant. Boyle showed that the volume V occupied by the air was inversely proportional to the pressure p of the air.

$$p \propto \frac{1}{V} \qquad \text{or} \qquad pV = \text{a constant}$$

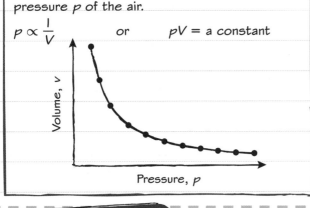

🧪 Practical skills — Demonstrating Boyle's law

Boyle's law can be demonstrated using the apparatus shown.

The air is trapped in the glass tube and its volume found from the scale. Pressure is transmitted to the gas by a column of oil. The pressure is changed using a foot pump, and its value read from the gauge.

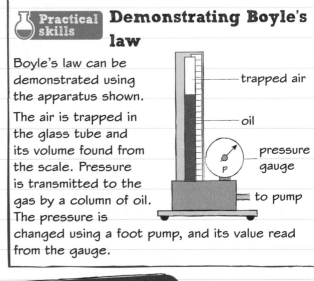

trapped air
oil
pressure gauge
to pump

Worked example

The table shows some data collected from an experiment using the apparatus shown above.

p(kPa)	V(10^{-6} m³)
130	33.4
155	27.9
220	19.8

Use the data to show that Boyle's law applies. **(3 marks)**

If Boyle's law applies pV should be constant.

130 × 33.4 = 4340; 155 × 27.9 = 4320;
220 × 19.8 = 4356

So pV is approximately constant and Boyle's law applies.

Worked example

145 cm³ of air at a pressure of 102 kPa in a bicycle pump is compressed to a volume of 85 cm³. Calculate the pressure of the compressed air in the pump. State the assumptions you must make.

(4 marks)

Unit conversions: 1 cm is 0.01 m, or 1 × 10⁻² m, so 1 cm³ is (1 × 10⁻² m)³ = 1 × 10⁻⁶ m³.

$$p_1 V_1 = p_2 V_2$$

$$p_2 = \frac{p_1 V_1}{V_2}$$

$$\therefore p_2 = \frac{102 \times 10^3 \times 145 \times 10^{-6}}{85 \times 10^{-6}}$$

$$= 1.74 \times 10^5 \text{ Pa}$$

The assumptions are that the temperature remains constant and the mass of air in the pump remains constant.

Now try this

A balloon rises above the ground. The volume of the balloon is 2 m³ at ground level where the pressure is 102 kPa. Calculate the volume of the balloon when it has risen to a height where the atmospheric pressure is 95 kPa. Assume the temperature remains constant.

Explain what will happen to the balloon as it continues to rise.

(4 marks)

The gas laws: the pressure law

The pressure exerted by a gas can be related to its temperature.

The pressure law

Joseph Louis Gay-Lussac investigated how the pressure exerted by a fixed mass of air depends on the temperature. In his experiment the volume occupied by the gas was kept constant.

The data indicates a linear, but not proportional, relationship between the pressure p exerted by the air and the Celsius temperature (the line on the graph does not pass through the origin).

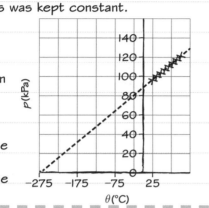

Demonstrating the pressure law

The pressure law can be demonstrated using the apparatus shown.

The temperature of the air trapped in the flask is changed by heating the water and the pressure exerted is measured using the pressure gauge.

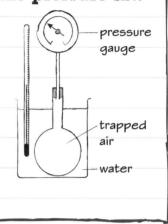

🧮 Maths skills The pressure law and absolute zero

If the graph above is extrapolated back to the point at which the pressure exerted by the air would be zero, it meets the temperature axis at about $-273\,°C$.

If we assign this temperature a value of 0 K, then we have a directly proportional relationship between pressure and temperature.

By adding 273 to each Celsius temperature we have converted the Celsius temperatures into absolute temperatures, T.

$$T = \theta + 273$$

The pressure law for absolute temperature

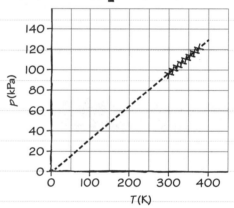

If the graph is now redrawn with kelvin instead of degrees Celsius on the temperature axis. It has a straight line through the origin, which indicates that

$$p \propto T \qquad \text{or} \qquad \frac{p}{T} = \text{constant}$$

Worked example

A gas in a sealed container exerts a pressure of 103 kPa at 25 °C. The container may leak if the internal pressure exceeds 135 kPa. Calculate the Celsius temperature at which the container will start to leak. **(2 marks)**

$\dfrac{p_1}{T_1} = \dfrac{p_2}{T_2}$ (assume volume stays constant)

$T_2 = \dfrac{p_2}{p_1} \times T_1 = \dfrac{135}{103} \times (25 + 273) = 390.6\,K$

$\theta = 390.6 - 273 = 118\,°C$

Both p_1 and p_2 are expressed in kPa, so there is no need to convert into Pa.

If two quantities are directly proportional, we can introduce a scaling constant and write an equation instead of a proportionality.

So $p \propto T$ becomes $p = aT$ (a is a constant)

$p_1 = aT_1 \qquad$ and $\qquad p_2 = aT_2$

Hence we can write $\dfrac{p_1}{T_1} = \dfrac{p_2}{T_2}$.

Now try this

1 When a fixed mass of air is heated at constant volume, the pressure exerted by the air increases as the temperature rises. Explain why the pressure exerted by the air increases, including ideas of momentum. **(4 marks)**

2 A gas in a sealed container of fixed volume exerts a pressure of 116 kPa at 40 °C. The container is cooled to 0 °C. Calculate the pressure of the gas after cooling. **(2 marks)**

The equation of state of an ideal gas

There is a simple relationship between pressure, volume and temperature for an ideal gas.

The general gas law

For a fixed mass of an ideal gas:

pV = constant as long as the temperature of the gas remains constant (Boyle's law).

$\dfrac{p}{T}$ = constant as long as the volume occupied by the gas remains constant (pressure law).

In addition, $\dfrac{V}{T}$ = constant as long as the pressure exerted by the gas remains constant (Charles' law).

We can sum these up as $\dfrac{pV}{T}$ = constant for a

fixed mass of an ideal gas: $\dfrac{p_1 V_1}{T_1} = \dfrac{p_2 V_2}{T_2}$

Equation of state of an ideal gas

The general gas law can be written $pV = KT$.

K depends upon the number of molecules in the gas. We therefore define it in terms of a fixed number of molecules, the mole.

A mole of gas contains Avogadro's number of molecules, N_A (6.02×10^{23} mol^{-1}).

For 1 mole of gas $pV = RT$, where R is the universal molar gas constant and equal to 8.31 J mol^{-1} K^{-1}.

For n moles of an ideal gas we can write

$pV = nRT$

This is the equation of state of an ideal gas.

Avogadro's law

Avogadro's law states that equal volumes of all gases at the same temperature and pressure contain the same number of molecules.

This is confirmed by the equation $pV = nRT$, which applies to any ideal gas.

$n = \dfrac{pV}{RT}$, so if p, V and T are the same, then so

is n (R is a constant for all gases).

What is a mole?

A mole is the amount of substance that contains the same number of particles as there are atoms in 12 g of the carbon-12 isotope.

1 mol N$_2$ 1 mol O$_2$

28 g   32 g

One mole of any ideal gas under given conditions of temperature and pressure will occupy the same volume as any other.

Worked example

A helium weather balloon has a volume of 7.4×10^4 m^3 at ground level where the pressure is 101 kPa. Calculate the new pressure exerted by the helium if the volume increases by 10% and the temperature changes from 20 °C to −5 °C as the balloon rises from ground level. **(2 marks)**

$\dfrac{p_1 V_1}{T_1} = \dfrac{p_2 V_2}{T_2}$

$p_2 = \dfrac{101 \times 7.4 \times 10^4 \times (273 - 5)}{(7.4 \times 10^4 + 0.74 \times 10^4) \times (273 + 20)}$

$= 84$ kPa

Worked example

Calculate the volume occupied by 1 mole of an ideal gas at standard temperature and pressure. **(2 marks)**

$V = \dfrac{nRT}{p} = \dfrac{1 \times 8.31 \times 273}{1.01 \times 10^5} = 0.0225$ m^3

Standard temperature is 0 °C and standard pressure is 1.01 × 10^5 Pa.

The pressure p_1 in the calculation was in kPa, so the answer must be in the same units.

Now try this

1 A student carries out an experiment to investigate how the volume occupied by a gas depends upon the temperature. The data obtained is plotted in the graph shown.

Explain how the experimental data may provide evidence for an absolute zero of temperature. **(3 marks)**

2 Calculate the number of moles of nitrogen gas in a 1000 cm³ flask under conditions of standard temperature and pressure. State any assumptions you have made. **(3 marks)**

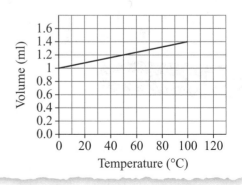

The kinetic theory equation

This equation relates the large-scale properties of a gas to the behaviour of molecules in the gas.

The kinetic theory equation

This equation is derived by assuming that the pressure of the gas, p, is a result of the collisions made between molecules and the walls of the container, of volume V, in which the gas is enclosed.

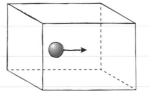

For simplicity a rectangular box is considered.

An expression for the force exerted by a single molecule, of mass m, on one side of the box is found, and then statistical methods are used to generalise this to take into account the effect of all the molecules, N, in the box.

This leads to the equation

$$pV = \tfrac{1}{3} N m \overline{c^2}$$

where $\overline{c^2}$ is the **mean square speed** of the gas molecules.

The mean square speed has units $m^2\,s^{-2}$, as it is the square of a speed.

> Note that mean square speed is not the same as the square of the mean of the speeds.

Molecular speeds

In the derivation of the kinetic theory equation, the mean square speed of the molecules is required.

This is the mean (average) of the squares of all the molecular speeds:

$$\overline{c^2} = \frac{c_1^2 + c_2^2 + c_3^2 + \dots + c_N^2}{N}$$

We take the **root mean square speed**, c_{rms}, as a measure of the 'average' speed of the molecules.

$$c_{rms} = \sqrt{\overline{c^2}} = \sqrt{\frac{c_1^2 + c_2^2 + c_3^2 + \dots + c_N^2}{N}}$$

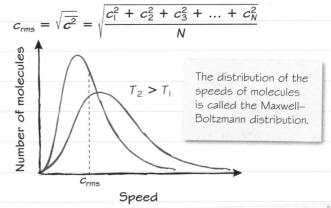

$T_2 > T_1$

> The distribution of the speeds of molecules is called the Maxwell–Boltzmann distribution.

> Note that the mean velocity of the molecules is zero.

Worked example

A balloon of volume $0.0145\,m^3$ contains $2.5\,g$ of helium gas at a pressure of $106\,kPa$. Calculate the r.m.s. speed of the helium atoms in the balloon.

(3 marks)

Total mass of gas = number of atoms N × mass of an atom m

$Nm = 2.5 \times 10^{-3}\,kg$

$$\overline{c^2} = \frac{3pV}{Nm} = \frac{3 \times 106 \times 10^3 \times 0.0145}{2.5 \times 10^{-3}}$$

$$= 1.84 \times 10^5\,m^2\,s^{-2}$$

$$\therefore c_{rms} = \sqrt{1.84 \times 10^6} = 1360\,m\,s^{-1}$$

Worked example

There are 5 molecules in a box. At one instant their speeds are $100\,m\,s^{-1}$, $200\,m\,s^{-1}$, $300\,m\,s^{-1}$, $400\,m\,s^{-1}$, and $500\,m\,s^{-1}$. Calculate: (a) the mean speed of the molecules; (b) the r.m.s. speed of the molecules.

(4 marks)

(a) $\overline{c} = \dfrac{100 + 200 + 300 + 400 + 500}{5}$

$= 300\,m\,s^{-1}$

(b) $c_{rms} =$

$$\sqrt{\frac{(100)^2 + (200)^2 + (300)^2 + (400)^2 + (500)^2}{5}}$$

$$\therefore c_{rms} = 332\,m\,s^{-1}$$

Now try this

1　A gas is maintained at a constant temperature. The volume occupied by the gas is reduced and the pressure exerted by the gas increases. By considering the motion of the molecules in the gas, explain why this increase in pressure is observed.

(3 marks)

2　The helium balloon in the example above is warmed to a temperature of $30\,°C$. Its volume ($0.0145\,m^3$) and the mass of gas it contains ($2.5\,g$) remain unchanged.
(a) Calculate the new pressure in the balloon.
(b) Calculate the new r.m.s. speed of the helium atoms in the balloon.

(3 marks)

The internal energy of a gas

The internal energy of an ideal gas is the sum of the kinetic energies of all the molecules in the gas.

Internal energy

Average molecular kinetic energy EK is given by the expression

$EK = \frac{1}{2}m\overline{c^2}$

where m is the mass of each molecule, and $\overline{c^2}$ is the mean square speed of the molecules (see page 91).

The internal energy U of an ideal gas is the sum of the molecular kinetic energies. This is because molecules in an ideal gas do not interact, except during collisions, so potential energy need not be considered). So, for N molecules

$U = \frac{1}{2}Nm\overline{c^2}$

$\overset{\bullet}{\underset{m}{}} \xrightarrow{\quad} c$

The internal energy of an ideal gas depends only on molecular motion

The meaning of absolute temperature

The ideal gas equation is $pV = nRT$.

Since $n = \dfrac{N}{N_A}$ $pV = N\left(\dfrac{R}{N_A}\right)T$

Also, $k = \dfrac{R}{N_A}$ so $pV = NkT$

Here k is the Boltzmann constant and N_A is the Avogadro constant.

$pV = NkT$ and $pV = \frac{1}{3}Nm\overline{c^2}$

See page 91 for the kinetic theory equation and page 90 for the ideal gas equation and the Avogadro constant.

So $NkT = \frac{1}{3}Nm\overline{c^2}$

$\therefore\ 3kT = m\overline{c^2}$

$\frac{3}{2}kT = \frac{1}{2}m\overline{c^2}$

Hence $T \propto \frac{1}{2}m\overline{c^2}$

This means that absolute (kelvin) temperature is directly proportional to the average kinetic energy of the molecules in an ideal gas.

Internal energy and temperature

T_1 $T_1 < T_2$ T_2

heating

$\frac{1}{2}m\overline{c^2} = \frac{3}{2}kT$

$\therefore\ U = N \times \frac{1}{2}m\overline{c^2} = \frac{3}{2}NkT$

$Nk = nR$

So $U = \frac{3}{2}RT$

Note that gases made up of molecules with more than one atom per molecule have more internal energy than $\frac{3}{2}RT$ because of rotational and vibrational energy associated with the molecules.

Worked example

A sample of air at a temperature of 27 °C is heated to 327 °C.

Calculate the ratio of the r.m.s. speed of the molecules in the container at 327 °C to the r.m.s. speed of the molecules at 27 °C assuming the air behaves as an ideal gas.   **(3 marks)**

$T \propto \frac{1}{2}m\overline{c^2}$ $\therefore\ \dfrac{\overline{c_2^2}}{\overline{c_1^2}} = \dfrac{T_2}{T_1}$

$\therefore\ \dfrac{\overline{c_2^2}}{\overline{c_1^2}} = \dfrac{327 + 273}{27 + 273} = \dfrac{600}{300} = 2$

$\therefore\ \dfrac{\overline{c_2^2}}{\overline{c_1^2}} = \sqrt{2}$

Now try this

A balloon filled with helium gas is left in direct sunlight. The temperature of the gas increases and hence the pressure exerted by the gas increases.

(a) By considering the motion of the molecules in the helium gas, explain why there is an increase in pressure.

(3 marks)

(b) The helium in the balloon is initially at a temperature of 18 °C. As the sunlight heats it, the r.m.s. speed of the helium atoms increases by 3%. Calculate the final temperature in °C of the helium in the balloon.   **(4 marks)**

Exam skills

This exam-style question uses knowledge and skills you have already revised. Have a look at page 35 for a reminder of Newton's laws of motion and at pages 87, 90, 91 and 92 for a reminder about gas pressure and the gas laws.

Worked example

A scuba diver breathes air from a gas cylinder. The volume of the cylinder is $0.0060\,\text{m}^3$ and the pressure of air inside the cylinder is $2.0 \times 10^7\,\text{Pa}$.

(a) Explain, in terms of molecules and Newton's laws of motion, how the air inside the cylinder exerts a pressure on its walls. **(5 marks)**

> Read the question carefully – you need to use a particle model to answer this question and you must refer to Newton's laws of motion. It is worth pausing before answering this question to plan how you will structure your answer.

The molecules are in rapid random motion. As they collide with the wall, the wall exerts a force on the molecules that changes their momentum. By Newton's third law, the molecules exert an equal but opposite force on the wall. By Newton's second law, the average force on the wall is equal to the average rate of change of momentum of the molecules colliding with the wall. This creates a pressure on the wall because pressure is equal to force per unit area.

(b) Explain in terms of molecular motion why the pressure falls if the temperature of the gas is reduced. **(3 marks)**

> It would not be sufficient here simply to state that the molecules move more slowly. A complete answer must link to the way they cause pressure (through rate of change of momentum). There are two related effects – the molecules collide with the walls less frequently *and* less violently. Both reduce the average pressure on the container walls.

The mean kinetic energy per molecule is proportional to the absolute temperature T. As T falls, the molecules move more slowly, so collisions are less violent and less frequent. This reduces the average rate of change of momentum at the wall and reduces the pressure.

(c) If all the air from the cylinder were extracted into a balloon free to expand in the atmosphere (at pressure $1.0 \times 10^5\,\text{Pa}$) with no change of temperature, what would the final volume of the balloon be? **(2 marks)**

> Note how the answer starts by quoting the relevant gas law: $pV = NkT$.

$pV = NkT$ so $V = \dfrac{NkT}{p}$

The only thing that changes is p, which falls by a factor of $\dfrac{2.0 \times 10^7}{1.0 \times 10^5} = 200$. The escaped air would occupy a volume $= 200 \times 0.0060 = 1.2\,\text{m}^3$.

Angular velocity

Objects moving in circular paths experience angular displacement and have an angular velocity.

Maths skills — Velocity around the circle

We can describe the speed of an object moving in a circle in terms of its velocity v along a straight line at a tangent to the circle (often called the linear or tangential velocity).

$$v = \frac{2\pi r}{T}$$

where r is the radius of the circle and T is the period, the time taken for one complete revolution. The direction of the velocity changes continuously.

Angular velocity

Instead of the changing linear velocity of an object moving in a circle, we can consider the object's angular displacement, $\Delta\theta$, measured in radians.

The angular velocity, ω, is given by

$$\omega = \frac{\Delta\theta}{\Delta t}$$

where $\Delta\theta$ is the angular displacement in a time Δt. Angular velocity has units rad s^{-1}.

In the time T for one complete revolution, the object will rotate through an angle of 2π radians and so we can write

$$\omega = \frac{2\pi}{T} \qquad \text{or} \qquad \omega = 2\pi f$$

where f is the frequency $\frac{1}{T}$, the number of revolutions in unit time.

Maths skills — Radians

The angle θ in radians is defined as $\theta = \frac{L}{r}$, where L is the arc length that gives the angle θ at the centre of the circle of radius r.

In a complete circle there are 2π radians (360°), so to convert an angle in degrees into radians we divide the angle by 360° and multiply by 2π.

Worked example

A vinyl record on a turntable rotates at 45 rpm.

(a) Calculate the time for one rotation of the disc.　**(1 mark)**

$$T = \frac{60}{45} = 1.33\,\text{s}$$

(b) Calculate the disc's angular velocity.　**(2 marks)**

$$\omega = \frac{2\pi}{T} = \frac{2\pi}{1.33} = 4.7\,\text{rad s}^{-1}$$

Worked example

Show that tangential velocity v and angular velocity ω are related by the expression $v = \omega r$.　**(2 marks)**

$$v = \frac{2\pi r}{T} = \frac{2\pi}{T} \times r$$

and $\omega = \dfrac{2\pi}{T}$

Hence $v = \omega r$.

Now try this

A playground roundabout is spun and takes 5 s to make one complete revolution.

(a) Calculate the angular velocity of the roundabout.　**(2 marks)**

A child is sitting close to the centre of the roundabout, and is rotating in a circle of radius of 1.2 m. Another child is holding on at the outside of the roundabout, which has a radius of 2.5 m.

(b) Compare the speeds of these two children.　**(2 marks)**

Centripetal force and acceleration

Objects moving in circular paths are accelerated towards the centre of the circle.

Centripetal force

An object moving in a circular path is always changing direction. According to Newton's first law, the forces acting on the object cannot be in equilibrium, so a resultant force must be acting on it.

Alternatively, we know that the object's velocity is changing, as velocity is a vector quantity. Hence the object is accelerating and a resultant force must act.

If the object moves with a constant speed, the resultant force must be directed towards the centre of the circular path.

This resultant force is called the **centripetal force**.

Note that the centripetal force is a name for a net force. It is the result of other forces acting on the object, and should not be thought of as a separate force.

Centripetal acceleration

The resultant force produces a centripetal acceleration, a. This acts towards the centre of the circle.

$$a = \frac{v^2}{r}$$

where r is the radius and v is the tangential velocity (the velocity around the circle).

Since $v = \omega r$ we can rewrite the expression for the centripetal acceleration as

$$a = \frac{(\omega r)^2}{r} = \omega^2 r$$

Calculating the centripetal force

To calculate the centripetal force we apply Newton's second law in the form $F = ma$.

$$F = \frac{mv^2}{r}$$

Since $a = \omega^2 r$, we can also calculate the centripetal force using
$F = m\omega^2 r$.

Constant speed in a circle

The acceleration is always directed towards O, the centre of the circle.

🧪 **Practical skills** **Investigating circular motion**

Tie a piece of string to a rubber bung of mass M, and then thread it through a short length of glass tubing. Attach a small mass m to the lower end of the string and make a mark on the string so that you can whirl the bung in circle of constant radius r. Use a metre rule to measure r, and a top-pan balance to measure M and m.

Whirl the bung round in a horizontal circle while holding the glass tube, keeping the radius of the bung's orbit constant. Using a stopwatch, measure the time t taken for 10 complete orbits of the bung.

The average orbital time T is calculated using $T = \frac{t}{10}$.

Repeat the experiment with different masses.

Calculate the velocity of the bung in the orbit using $v = \frac{2\pi r}{T}$ and then work out v^2.

— rubber bung

— glass tubing

← r →

— mass holder and masses

Plot a graph of v^2 against m.

If friction is negligible, the resultant force on the bung is equal to the weight of m.

Applying Newton's second law, $\frac{Mv^2}{r} = mg$

so $v^2 \propto m$. Hence the graph should be a straight line passing through the origin.

Worked example

A car of mass 1650 kg goes around a roundabout in a circle of diameter 27.5 m at a constant speed of 12.0 m s⁻¹. Calculate the resultant force acting on the car and state the origin of this force. **(3 marks)**

$$F = \frac{mv^2}{r} = \frac{1650 \times (12)^2}{27.5/2} = 17\,300\,N$$

The force arises from the frictional force between the car tyres and the road surface.

Now try this

The London Eye is a giant wheel with a diameter of 120 m. It makes one complete revolution every 30 minutes. Calculate the centripetal acceleration of a passenger on the wheel. **(3 marks)**

Simple harmonic motion

Simple harmonic motion is a form of periodic motion in which the acceleration is proportional to the displacement and in a direction opposite to the displacement.

Periodic motion

There are many examples of **periodic motion**.

The Moon orbiting the Earth, the pendulum in a pendulum clock, and a child on a swing are all examples of periodic motion.

All have a repeating motion in which the time period for each repeat is roughly constant.

As a young man in 16th century Pisa, Galileo made an important observation. He noticed that the chandeliers in church swung back and forth with a time period that was approximately constant.

Galileo Galilei

Simple harmonic motion

If the time taken by an object in periodic motion to **oscillate** (move back and forth) is constant, even if the amplitude of oscillation changes, the oscillation is described as **isochronous**.

We call these oscillating systems simple harmonic oscillators. We say that they exhibit **simple harmonic motion** (s.h.m.).

Restoring force

For a body to exhibit s.h.m. a simple condition must be met.

When the body is displaced from equilibrium, there must be a restoring force, F, directed towards the equilibrium point, which produces an acceleration in that direction which is directly proportional to the displacement, x, of the body from the equilibrium point.

The condition for s.h.m.

$a \propto -x$

$a = -\omega^2 x$ where ω is the angular frequency.
A force F acts to bring the body to a position where the forces are in equilibrium. The force can be provided by a spring that obeys **Hooke's law**:
$F = -kx$

Mass–spring systems that obey Hooke's Law are described as examples of **exact s.h.m.**

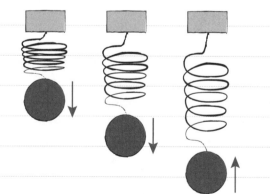

Approximate s.h.m.

A mass suspended by a light string swings freely when displaced and released.

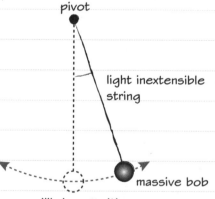

light inextensible string

massive bob

equilibrium position

The pendulum swings back and forth about its equilibrium position. The mass experiences a restoring force (a component of its weight) that accelerates it back to its equilibrium position, but this force is only approximately proportional to the displacement.

Now try this

(a) State the conditions that must be met for an object to move with s.h.m.

(b) Making use of a free-body force diagram, explain why the motion of a simple pendulum can only be considered to approximate to s.h.m.

(5 marks)

Solving the s.h.m. equation

The solution to the s.h.m. equation $a = -\omega^2 x$ can be obtained from a comparison between s.h.m. and circular motion.

Circular motion and s.h.m.

Both circular motion and s.h.m. are examples of a regularly repeating motion, and there are closer similarities between them.

This can be pictured by setting up a mass–spring system and a pin on a vertical turntable, and matching their periods – the mass and the pin are always at the same height.

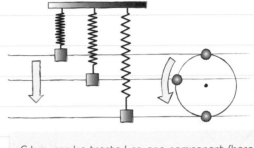

S.h.m. can be treated as one component (here, the vertical one) of uniform circular motion.

Simple harmonic motion solved

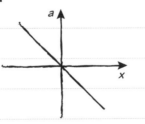

Point P moves around the circle with a constant angular velocity $\omega \left(= \dfrac{2\pi}{T}\right.$, where T is the period$)$.

The projection of P onto the diameter of the circle is C and moves between A and B.

The displacement x of C from O is given by

$x = r \cos\theta$

But $\theta = \omega t$, so $x = r \cos\omega t$

The velocity v of C is given by $v = -\omega r \sin\omega t$

The acceleration of C is given by $a = -\omega^2 r \cos\omega t = -\omega^2 x$

Hence C moves with simple harmonic motion.

See page 98 for the relationship between displacement, velocity and acceleration in s.h.m.

Acceleration and displacement

The condition for s.h.m. is

$a = -\omega^2 x$

Acceleration a is always in the opposite direction to displacement from the equilibrium position x, hence the negative gradient.

For a mass m on a spring, $F = -kx$ where k is the spring constant, and $F = ma$.

Hence $a = -\left(\dfrac{k}{m}\right)x$

Therefore $\omega^2 = \dfrac{k}{m}$

$\dfrac{2\pi}{T} = \sqrt{\dfrac{k}{m}} \qquad \therefore T = 2\pi\sqrt{\dfrac{m}{k}}$

Worked example

A child's toy consists of a ball of mass 50 g attached to a long spring. When hung from a support and displaced vertically, the toy oscillates in s.h.m. with a time period of 0.75 s. Calculate the spring constant of the spring. **(2 marks)**

$T = 2\pi\sqrt{\dfrac{m}{k}} \quad \therefore \; 0.75 = 2\pi\sqrt{\dfrac{0.05}{k}}$

$k = \dfrac{4\pi^2 \times 0.05}{(0.75)^2} = 3.5\,\text{N m}^{-1}$

Now try this

A car of mass 1200 kg has a suspension system that consists of a set of four springs with a combined spring constant of 30 kN m^{-1}. The car is set into oscillation with a driver of mass 80 kg sitting in it. Calculate the period of oscillation of the car. **(4 marks)**

Graphical treatment of s.h.m.

For an object oscillating with s.h.m., the displacement, velocity and acceleration all vary sinusoidally, but with phase differences.

S.h.m. graphs

We can plot displacement x against time t and then plot velocity v and acceleration a against t from the gradient of the first graph.

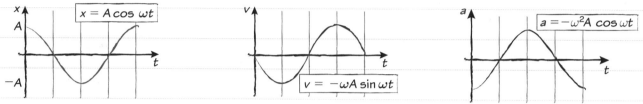

$x = A \cos \omega t$

$v = -\omega A \sin \omega t$

$a = -\omega^2 A \cos \omega t$

Phase difference

Each quantity – displacement, velocity and acceleration – varies sinusoidally, but with a **phase difference** between the quantities. This means that each quantity varies in the same way (and with the same frequency), but the times at which each quantity is zero are different. As a result, each graph has the same shape, but the graphs are displaced along the time axis with respect to each other. Velocity is $\frac{\pi}{2}$ rad behind displacement, and acceleration is $\frac{\pi}{2}$ rad behind velocity.

If displacement is a maximum at time $t = 0$ (for example, if a mass on the end of a stretched spring is released), then we write displacement as a cosine function, $x = A \cos\omega t$.

If initial displacement is zero (for example, if a pendulum bob is flicked away from its rest position), then we write it as a sine function, $x = A \sin\omega t$.

Practical skills — Simple harmonic oscillation

$f = \frac{1}{T}$ where f is the frequency and T is the period for one complete oscillation in seconds.

The angular frequency ω, in rad s^{-1}, is thus

$\omega = \frac{2\pi}{T}$ or $\omega = 2\pi f$.

To determine the period/frequency of a simple harmonic oscillator, set up a mass–spring system or a pendulum as shown and time a number of oscillations.

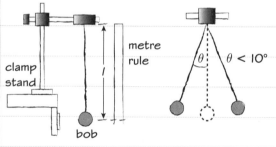

clamp stand

metre rule

l

bob

θ $\theta < 10°$

- The periodic time for a pendulum is constant only when amplitudes are small.
- Measure many oscillations to calculate the average time for one oscillation.
- Increase the total time measured.
- Make timings by sighting the mass past a fixed reference point.
- The mass moves fastest at the equilibrium position and slowest at the maximum displacement positions.

Worked example

A mass is moving with s.h.m. with amplitude A and angular frequency ω. Its displacement is given by $x = A \sin\omega t$, and its velocity by $v = \omega A \cos\omega t$. Show that its velocity at a displacement x is given by $v = \pm\omega\sqrt{A^2 - x^2}$. **(3 marks)**

$v = \omega A \cos\omega t$ $\therefore$ $v^2 = \omega^2 A^2 \cos^2 \omega t$

$\therefore$ $v^2 = \omega^2 A^2 (1 - \sin^2 \omega t)$

$x^2 = A^2 \sin^2 \omega t$ $\therefore$ $v^2 = \omega^2 A^2 \left(1 - \frac{x^2}{A^2}\right)$

$\therefore$ $v^2 = \omega^2 (A^2 - x^2)$ $\therefore$ $v = \pm\omega\sqrt{A^2 - x^2}$

(Note that it follows that $v_{max} = \omega A$.)

Now try this

A mass is hung from the end of a spring and set into vertical oscillation. The amplitude of oscillation is 4.5 cm, and the maximum acceleration of the mass is 2.2 cm s^{-2}. Calculate the maximum velocity of the mass. **(4 marks)**

Energy in s.h.m.

Potential and kinetic energy are continuously interchanged when an object moves in simple harmonic motion.

Energy terms in s.h.m.

Consider a mass on a spring.

Displacement $x = A\cos\omega t$

Velocity $v = -\omega A \sin\omega t$

Thus, kinetic energy

$E_K = \frac{1}{2}mv^2 = \frac{1}{2}m\omega^2 A^2 \sin^2\omega t$

and potential energy

$E_P = \frac{1}{2}kx^2 = \frac{1}{2}kA^2 \cos^2\omega t$

$\frac{k}{m} = \omega^2$

Therefore,

$E_P = \frac{1}{2}m\omega^2 A^2 \cos^2\omega t$

See page 97 for a reminder of the derivation of $\frac{k}{m} = \omega^2$.

Kinetic and potential energy transfer during oscillation

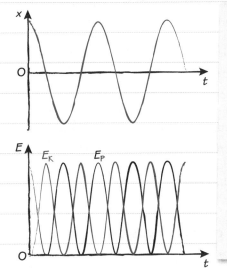

The object's kinetic energy is greatest when it passes through the equilibrium position at its maximum velocity; its potential energy is greatest when it reaches its maximum amplitude and stops for an instant as it changes direction.

Sum of kinetic and potential energy

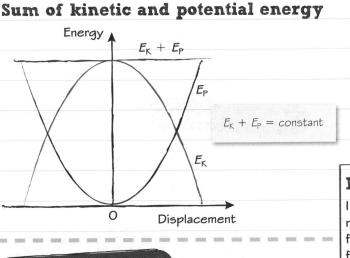

$E_K + E_P = $ constant

$E_K = \frac{1}{2}m\omega^2 A^2 \sin^2\omega t$

$E_P = \frac{1}{2}m\omega^2 A^2 \cos^2\omega t$

$E_K + E_P = \frac{1}{2}m\omega^2 A^2 (\sin^2\omega t + \cos^2\omega t)$

$\qquad = \frac{1}{2}m\omega^2 A^2 = $ constant

Damping

In real oscillating systems the total energy does not stay constant over time. Energy is removed from the oscillating system (e.g. by frictional forces) and the amplitude decreases over time. We say that the system is damped.

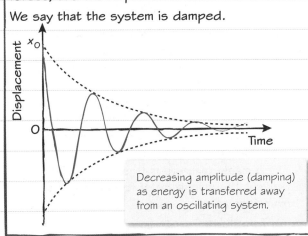

Decreasing amplitude (damping) as energy is transferred away from an oscillating system.

Worked example

A mass of 25 g is oscillating with s.h.m. with a time period of 3.4 s and amplitude of 5.0 cm.

Calculate the maximum kinetic energy of the mass. **(3 marks)**

$\omega = \frac{2\pi}{T} = \frac{2\pi}{3.4} = 1.85 \,\text{rad s}^{-1}$

$KE_{max} = \frac{1}{2}m\omega^2 A^2 = \frac{1}{2} \times 25 \times 10^{-3} \times (1.85)^2$

$\qquad\qquad \times (5 \times 10^{-2})^2$

$\qquad\qquad = 1.1 \times 10^{-4}\,\text{J}$

Now try this

Calculate the ratio of the kinetic energy of a simple harmonic oscillator to its potential energy when the displacement is half its amplitude. **(3 marks)**

Forced oscillations and resonance

Driving a system into oscillation at its natural frequency results in effective energy transfer.

Making things vibrate

If you push a swing once and then leave it, it swings (oscillates) at its **natural frequency**. When a mechanical system is set into oscillation and no external forces act on the system, like the swing, it is said to be a **free oscillation**.

If external forces act then energy is transferred.

Applying a force to drive the system into oscillation transfers energy from the **driver** to the oscillating system. This is a **forced oscillation**.

If the system is driven at its natural frequency there is an efficient transfer of energy, and the amplitude of the system increases over time.

Changing the driving frequency

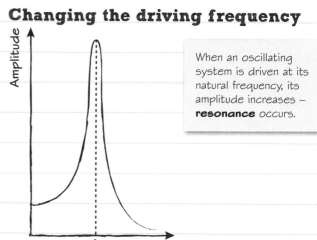

When an oscillating system is driven at its natural frequency, its amplitude increases – **resonance** occurs.

Damping of driven systems

When energy is removed from an oscillating system (damping), the response of the system is reduced when it is forced into oscillation. The maximum amplitude reduces, as does the frequency at which resonance occurs.

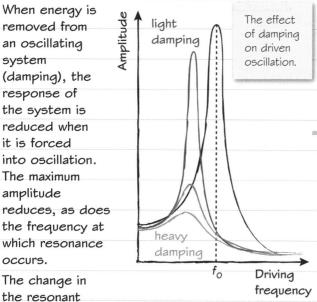

The effect of damping on driven oscillation.

The change in the resonant frequency due to damping is only small.

Examples of forced and damped systems include:

- Felt dampers rest on the strings in a piano and prevent the strings from vibrating once the pianist's fingers leave the keys.
- Shock absorbers in a car are designed to remove vibrational energy from the car as it is forced into vertical motion by changes in the road surface.
- Tuned mass dampers in bridges and buildings prevent the amplitude of forced oscillations from increasing.

Unwanted resonance

The swaying of a tall building caused by the wind, the oscillation of a bridge as pedestrians walk across it, the vibration of a loudspeaker enclosure when particular notes are sounded are all examples of forced oscillations. If the forcing frequency matches the natural frequency of the driven system, then unwanted resonance occurs.

Worked example

A car of mass 1150 kg has a suspension system of effective spring constant 1.02×10^5 N m^{-1}. Calculate the frequency at which resonance will occur.

(3 marks)

$$T = 2\pi\sqrt{\frac{m}{k}} = 2\pi \times \sqrt{\frac{1150}{1.02 \times 10^5}} = 0.667\,\text{s}$$

$$\therefore f = \frac{1}{T} = \frac{1}{0.667} = 1.50\,\text{Hz}$$

Maths skills   Rounding errors from the intermediate stages of a calculation can affect the final answer. Always retain enough significant figures in any intermediate answers that you work out, so that your final answer will be accurate.

Quote your final answer to no more than the number of significant figures in your least precise piece of data.

Now try this

The suspension system of a car is damped to give the passengers a smooth ride. Comment on whether the suspension system should be lightly or heavily damped.

(3 marks)

Exam skills

A car suspension system can be modelled as a damped mass–spring oscillator, as shown in the diagram.

— spring, spring constant k

— mass m

The fluid damper provides a frictional force that opposes the motion of the oscillator.

— fluid damper

(a) (i) When the car goes over a bump in the road it oscillates vertically with a time period of 0.80 s and an initial amplitude of 10 cm. The car, with driver, has a total mass of 1250 kg. Calculate the effective spring constant of the car's suspension. **(2 marks)**

For a mass–spring system $T = 2\pi\sqrt{\dfrac{m}{k}}$

so $k = \dfrac{4\pi^2 m}{T^2} = 77\,000\,\text{N m}^{-1}$ (2 s.f.)

(ii) Sketch a graph to show how the displacement of the car varies with time from the moment it first reaches the bump in the road. Indicate approximate values on the axes. **(3 marks)**

(iii) Calculate the maximum energy stored in the oscillation after the car hits the bump. **(3 marks)**

For simple harmonic motion: $v = -\omega A\sin\omega t$

Kinetic energy $E_K = \dfrac{1}{2}mv^2 = \dfrac{1}{2}m\omega^2 A^2\sin^2\omega t$

$KE_{max} = \dfrac{1}{2}m\omega^2 A^2 = \dfrac{1}{2}m(2\pi T)^2 A^2 = 386\,\text{J}$

(iv) Explain what happens to this energy. **(2 marks)**

The oscillator does work against viscous damping forces, transferring the kinetic energy of the oscillator into heat in the fluid.

(b) When the car passes over a section of road with regular ridges it is forced to vibrate vertically. The amplitude of these vibrations is particularly large when the car is travelling at 18 m s⁻¹.

(i) Explain why the amplitude is particularly large at one particular speed. **(2 marks)**

At this speed the frequency at which the car encounters the ridges is equal to the natural frequency of vertical oscillations of the car, so the suspension system resonates at this speed.

(ii) Calculate the separation of the ridges on the road. **(3 marks)**

The time to pass from one ridge to the next equals the time period of the oscillator (0.80 s).

distance = vt = 18 × 0.80 = 14 m.

This exam-style question uses knowledge and skills you have already revised. Have a look at pages 96–100 for a reminder about simple harmonic motion.

A question like this one places physics in context, in this case modelling a car's suspension. You are not expected to know anything in detail about how cars are constructed or how their suspension is configured. You need to apply your knowledge of the important underlying physics – mass–spring systems, damping, forced oscillations and resonance.

Answers should be rounded to an appropriate number of significant figures. The data used in the calculation is to a minimum of 2 s.f. (10 cm and 0.80 s) so this answer has been rounded to 2 s.f. as well.

The important point here is to show that the oscillation is damped. The amount of damping doesn't matter, just make sure the amplitude decays. You are also asked to indicate values on the axes. The maximum displacement is 10 cm and the period of oscillation is 0.80 s so these should be marked on the axes. Make sure that the time period is shown after one complete cycle (or mark 0.40 s after half a cycle).

Notice how the working has been shown in logical steps, starting with the equation for velocity and the definition of kinetic energy.

Command word: Explain

If a question asks you to **explain** why something happens you should:

✓ write in full sentences

✓ use correct scientific language.

The calculation here follows from the explanation in part (i). The answer has also been rounded to 2 s.f.

Gravitational fields

Any object with mass has a gravitational field around it. The interaction of two or more gravitational fields produces gravitational forces.

Gravitational force between two masses

The gravitational force between two masses is always attractive.

A force of equal magnitude acts on each mass. For each pair of interacting masses the forces form an action–reaction pair (Newton's third law).

Mathematically, a spherical object can be treated like a point mass located at its centre.

Gravitational field at Earth's surface

Over a very small section of the sphere, the surface appears to be flat.

The field lines appear parallel, and we say that the field is approximately uniform.

This is the situation near the Earth's surface. The gravitational field is uniform and directed towards the Earth's surface.

Gravitational field around a spherical mass

Gravitational fields can be represented visually by field lines. The lines have direction, and the closer together they are, the stronger the field.

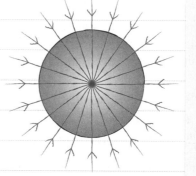

The field is directed radially inwards towards the centre of the object.
The radial field lines appear less divergent over a small section of the sphere.

A field in physics gives rise to a force – think of electric and magnetic fields.

Acceleration of free fall

Rearranging the field strength equation gives

$F = mg$

where F is the resultant force acting on an object in free fall at the point being considered.

Applying $F = ma$ (from Newton's second law) we can see that $mg = ma$.

Hence g is equal to the acceleration of free fall at a point. Alternative units for g are $m\,s^{-2}$.

Strength of a gravitational field

The strength of the gravitational field g at a point is the gravitational force F acting per unit mass on a small test mass m placed at the point.

$g = \dfrac{F}{m}$ g has units $N\,kg^{-1}$.

Note that g is a vector quantity – it has a magnitude and a direction. We must use vector addition to calculate the gravitational field strength at a point due to more than one mass.

Worked example

The first human flight to Mars is planned for 2026. A man of weight 755 N on the Earth would weigh only 295 N on Mars. Calculate the gravitational field strength at the surface of Mars.
(g at Earth's surface = 9.81 N kg⁻¹)   **(2 marks)**

$g = \dfrac{F}{m}$ $\therefore \dfrac{g_{Mars}}{g_{Earth}} = \dfrac{F_{Mars}}{F_{Earth}}$

$\therefore g_{Mars} = \dfrac{295}{755} \times 9.81 = 3.83\,N\,kg^{-1}$

Maths skills Instead of calculating the mass of the man and then using this to calculate a value for g, it is appropriate to use a ratio method. The mass of the man is a constant, and so the ratio of field strength values is equal to the ratio of gravitational force values.

Now try this

1 Show that $N\,kg^{-1}$ and $m\,s^{-2}$ are equivalent units.   **(1 mark)**

2 Using the gravitational field strength on the surface of Mars from the worked example above, calculate the weights on Mars of a Mars explorer who weighs 730 N on Earth, and a life-support pack, which has a mass of 35 kg on Earth.   **(3 marks)**

Newton's law of gravitation

The force between two masses, such as planets, obeys an inverse square law with distance.

The inverse square law for gravitational force

The force between two point masses decreases as the square of the distance between them:

$$F = -\frac{Gm_1m_2}{r^2}$$

where m_1 and m_2 are two point masses, r is their separation and G is the universal gravitational constant. The minus sign shows that the force is always attractive.

The gravitational force between two spherical masses

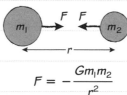

$$F = -\frac{Gm_1m_2}{r^2}$$

$$G = -6.67 \times 10^{-11}\,\text{Nm}^2\,\text{kg}^{-2}$$

The gravitational force due to a spherical mass is the same as that due to a point mass located at its centre.

The inverse square law for g

The gravitational field strength, g, of a uniform spherical mass behaves at any point at least one radius away from the centre of the sphere (that is, outside the sphere) like the field of a point mass positioned at the centre of the sphere.

Since g at a point is the force per unit mass placed at that point,

$$g = \frac{F}{m}$$

for a radial field produced by a point mass M

$$g = -\frac{GM}{r^2}$$

$$g \propto \frac{1}{r^2}$$

– that is, an inverse square relationship.

Graph of gravitational field strength for a spherical mass

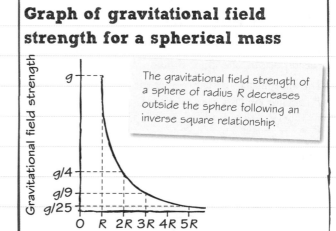

The gravitational field strength of a sphere of radius R decreases outside the sphere following an inverse square relationship.

Worked example

Europa and Ganymede are moons of Jupiter. Ganymede has 3.1 times the mass of Europa and 1.7 times Europa's diameter.

Calculate the ratio of the gravitational field strength at the surface of Ganymede to that at the surface of Europa. **(2 marks)**

$$g = -\frac{GM}{r^2} \quad \therefore \frac{g_G}{g_E} = \frac{M_G}{M_E} \times \frac{r_E^2}{r_G^2}$$

$$\therefore \frac{g_G}{g_E} = 3.1 \times \left(\frac{1}{1.7}\right)^2 = 1.1$$

Maths skills Take care with inverse relationships when finding ratios: check that you have placed the correct quantities in the denominator and the numerator.

Maths skills Also watch out for quantities that are raised to powers. It is easy to forget to apply the power when carrying out the calculation.

Now try this

How far would you have to travel upward from the Earth's surface to notice a 0.1 N kg⁻¹ difference in the Earth's gravitational field strength? Use your answer to justify the statement that the gravitational field of the Earth is uniform close to its surface. (Take the radius of the Earth as 6370 km and g at the Earth's surface as 9.81 N kg⁻¹.) **(6 marks)**

Kepler's laws for planetary orbits

Kepler formulated three laws for planetary motion in the 17th century.

Kepler's laws

Law 1: Planets orbit the Sun in an elliptical path with the Sun at one focus of the ellipse.

Law 2: An imaginary line between a planet and the Sun sweeps out equal areas during equal intervals of time.

Law 3: The square of the orbital period, T, of a planet is proportional to the cube of the average distance, r, of the planet from the Sun.

$T^2 \propto r^3$

Kepler's second law

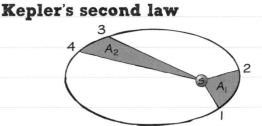

The two areas A_1 and A_2 represent areas swept out in equal time intervals. The planet travels faster between points 1 and 2 than it does between points 3 and 4, so $A_1 = A_2$.

Kepler's third law

Consider a planet of mass m in a circular orbit of radius about the Sun of mass M.

The magnitude of the resultant centripetal force, F, acting on the planet due to gravity is given by

$$F = \frac{GMm}{r^2}$$

Applying Newton's second law, $F = ma$, and the equation for centripetal acceleration, $a = \omega^2 r$

$$\frac{GMm}{r^2} = m\omega^2 r$$

$$\therefore \frac{GM}{r^2} = \left(\frac{2\pi}{T}\right)^2 r$$

$$T^2 = \left(\frac{4\pi^2}{GM}\right)r^3$$

Make sure you know this derivation.

Kepler's third law for our Solar System

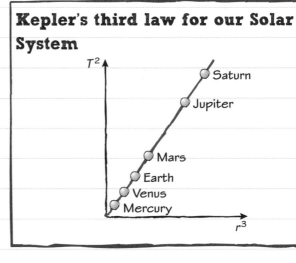

Worked example

The moons orbiting Jupiter follow the same laws of motion as the planets orbiting the Sun. Io has an orbital radius of 422 Mm and an orbital period of 1.77 (Earth) days. Ganymede has an orbital radius of 1070 Mm.

Calculate the period of Ganymede as it orbits Jupiter. **(2 marks)**

$$\frac{T_G^2}{T_I^2} = \frac{r_G^3}{r_I^3} \quad \therefore T_G = \sqrt{\left(\frac{1070}{422}\right)^3 \times (1.77)^2}$$

$$T_G = 7.15 \text{ days}$$

Maths skills When using ratios of quantities we do not have to use S.I. units as long as the quantities are expressed in the same units. So it is not necessary to convert the orbital period from days to seconds, nor to include the 10^6 factor to convert Mm to m.

Be careful when calculating quantities with powers and roots. If in doubt, break the calculation up into stages.

Now try this

The average orbital distance of Mars about the Sun is 1.52 times the average orbital distance of the Earth about the Sun. Use Kepler's third law to calculate the time taken for Mars to orbit the Sun. **(2 marks)**

Satellite orbits

The planets orbit the Sun in slightly elliptical orbits. However, many satellites have orbits around Earth that are circular or nearly circular.

Speed of a satellite

Consider a satellite of mass m in a circular orbit of radius r about the Earth of mass M.
The satellite is maintained in its orbit by the gravitational pull of the Earth, F.

$$F = -\frac{GMm}{r^2}$$

We can treat a satellite orbiting the Earth in the same way as we treated a planet orbiting the Sun on page 104.

This centripetal force is a resultant force, and so we can apply Newton's second law ($F = ma$) and then find speed v from $a = \frac{v^2}{r}$

$$\frac{GMm}{r^2} = \frac{mv^2}{r} \qquad \therefore v = \sqrt{\frac{GM}{r}}$$

Speed and orbital radius

$$v = \propto \frac{1}{\sqrt{r}}$$

In other words, the higher the satellite (the larger the radius of its orbit), the slower the satellite travels as it orbits the Earth.

A satellite in a low-level orbit (just above the atmosphere) has quite a large velocity – it takes less than 100 minutes to make one orbit.

A satellite 35 800 km above the Earth's surface will have an orbital period of 24 hours. If it orbits over the equator in the direction of rotation of the Earth, it will always be above the same point on the Earth. This is a **geostationary orbit**.

Geostationary orbit

Geostationary satellites are used for satellite TV and other types of global telecommunication.

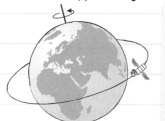

If a satellite has an orbital period of 24 hours and orbits over the equator in the direction of rotation of the Earth, it remains above the same point in the Earth's surface all the time.

Geosynchronous orbit

A satellite a **geosynchronous orbit** has a period of 24 hours but does not orbit above the equator.

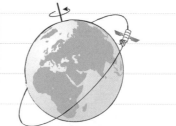

If a satellite has an orbital period of 24 hours, it passes over each point under its orbit at the same time every day.

Worked example

Calculate the orbital radius of a satellite for it to be in a geostationary orbit about the Earth. (The mass of the Earth is 6.0×10^{24} kg; $G = 6.67 \times 10^{-11}$ N m^2 kg^{-2}.) **(3 marks)**

$$T^2 = \left(\frac{4\pi^2}{GM}\right)r^3$$

$$r = \sqrt[3]{\frac{6.67 \times 10^{-11} \times 6.0 \times 10^{24} \times (24 \times 60 \times 60)^2}{4\pi^2}}$$

$$\therefore r = 4.2 \times 10^7 \text{ m}$$

Worked example

The planet Mercury has a mass of 3.30×10^{23} kg and a radius of 2.44×10^6 m. In 2011, the Messenger spacecraft became the first spacecraft to orbit Mercury. Messenger's orbital height varied. At its closest it was 201 km above Mercury's surface.

Calculate its orbital speed at this height. **(2 marks)**

$$v = \sqrt{\frac{GM}{r}}$$

$$= \sqrt{\frac{6.67 \times 10^{-11} \times 3.3 \times 10^{23}}{2.44 \times 10^6 + 2.01 \times 10^5}}$$

$$= 2900 \text{ m s}^{-1}$$

Now try this

In a Mars mission, a spacecraft will drop a rover onto the surface of Mars. The spacecraft will then take up a geostationary orbit above the rover. Calculate the orbital radius of the satellite and its height above the Martian surface. (The mass of Mars is 6.4×10^{23} kg, its radius is 3400 km and the length of its day is 24 hours and 37 minutes; $G = 6.67 \times 10^{-11}$ N m^2 kg^{-2}.) **(4 marks)**

Gravitational potential

Gravitational potential at a point is the work done in bringing unit mass from infinity to that point.

Gravitational potential

The **gravitational potential**, V_g, at a point in a gravitational field is defined as being the work done, W, when a 1 kg mass is brought from infinity to the point.

If the mass being moved is m we can write

$$V_g = \frac{W}{m}$$

V_g has units of $J\,kg^{-1}$.

Consider m initially an infinite distance away from M and being brought to a point a distance r from M.

The work done on m is given by

$$W = -\frac{GMm}{r}$$

The gravitational potential at r is therefore given by

$$V_g = \frac{W}{m} = -\frac{GM}{r}$$

Equipotentials

−30 MJ kg⁻¹
−40 MJ kg⁻¹
−50 MJ kg⁻¹

Each **equipotential** is a line joining points at which the gravitational potential is the same.

Work in a gravitational field

The shaded area represents the work done, W, in moving the mass m between x_1 and x_2.

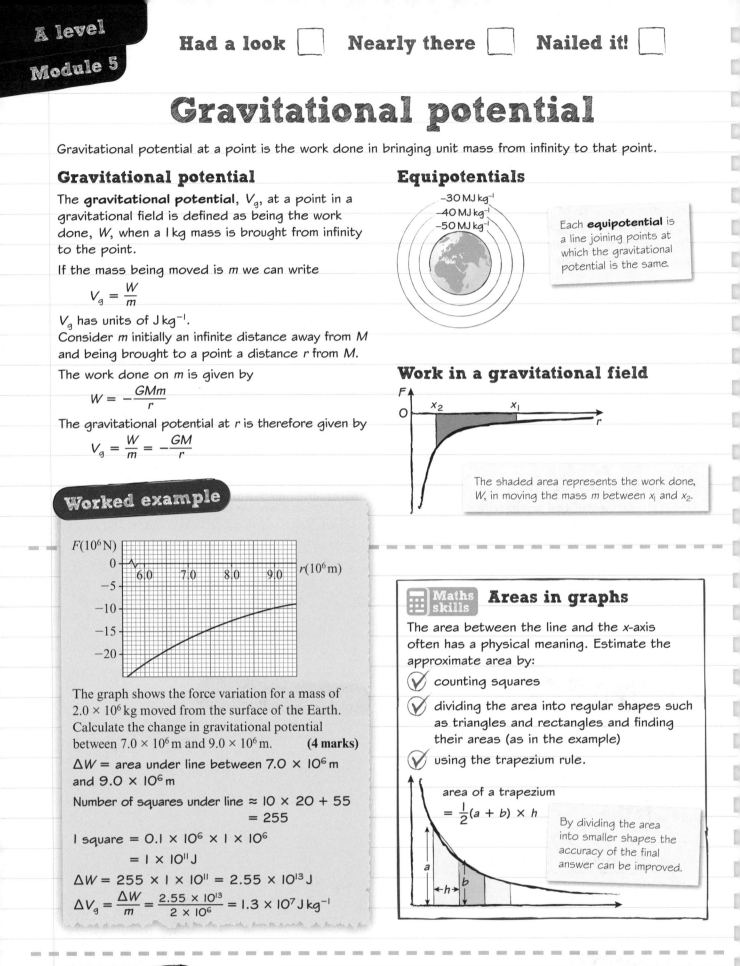

Worked example

The graph shows the force variation for a mass of 2.0×10^6 kg moved from the surface of the Earth. Calculate the change in gravitational potential between 7.0×10^6 m and 9.0×10^6 m. **(4 marks)**

ΔW = area under line between 7.0×10^6 m and 9.0×10^6 m

Number of squares under line ≈ $10 \times 20 + 55$
$= 255$

1 square $= 0.1 \times 10^6 \times 1 \times 10^6$
$= 1 \times 10^{11}$ J

$\Delta W = 255 \times 1 \times 10^{11} = 2.55 \times 10^{13}$ J

$\Delta V_g = \frac{\Delta W}{m} = \frac{2.55 \times 10^{13}}{2 \times 10^6} = 1.3 \times 10^7\,J\,kg^{-1}$

Maths skills — Areas in graphs

The area between the line and the x-axis often has a physical meaning. Estimate the approximate area by:

✓ counting squares

✓ dividing the area into regular shapes such as triangles and rectangles and finding their areas (as in the example)

✓ using the trapezium rule.

area of a trapezium
$= \frac{1}{2}(a + b) \times h$

By dividing the area into smaller shapes the accuracy of the final answer can be improved.

Now try this

Geostationary satellites move in an equatorial orbit at a height of 35 800 km above sea level. Calculate the change in gravitational potential when a satellite is placed in such an orbit from sea level. (Radius of the Earth is 6360 km; mass of the Earth is 5.98×10^{24} kg; $G = 6.67 \times 10^{-11}$ N m² kg⁻².) **(3 marks)**

Gravitational potential energy and escape velocity

Energy transfer may take place when masses move in a gravitational field.

Gravitational potential energy

The work done ΔW when a mass m is moved between two points is given by $\Delta W = m\Delta V_g$.

$$\Delta V_g = -\frac{GM}{r_2} - \left(-\frac{GM}{r_1}\right) \quad \therefore \quad \Delta W = GMm\left(\frac{1}{r_1} - \frac{1}{r_2}\right)$$

If the mass is brought from infinity, where the gravitational potential is 0, to r, the work done is

$$W = -\frac{GMm}{r}$$

We call this work the **gravitational potential energy** (E_{gp} or GPE).

$$E_{gp} = -\frac{GMm}{r}$$

This is negative, showing that work must be done on the mass to take it from r to infinity.

$$\Delta E_{gp} = GMm\left(\frac{1}{r_1} - \frac{1}{r_2}\right)$$

Escape velocity

If a spacecraft is intended to escape from the Earth's gravitational field altogether, it must have at least enough kinetic energy to match the change in gravitational potential energy in being displaced to infinity.

$$E_k \geqslant \Delta E_{gp}$$
$$\frac{1}{2}mv_{min}^2 = \frac{GMm}{r} \quad \therefore \quad v_{min} = \sqrt{\frac{2GM}{r}}$$

v_{min} is the **escape velocity**, the minimum launch velocity that a small mass in a gravitational field must have in order to move it to a point at infinity, that is, the velocity at which it will never fall back to Earth.

The escape velocity from Earth (for any object) is $1.1 \times 10^4\,\mathrm{m\,s^{-1}}$. By comparing this with the root mean square speeds of gas molecules in the atmosphere you can predict which gases will escape from the atmosphere into space over time.

Worked example

Show that for displacements near to the surface of the Earth the change in gravitational potential energy $\Delta E_{gp} = mg\Delta h$. **(3 marks)**

$$\Delta E_{gp} = GMm\left(\frac{1}{r_1} - \frac{1}{r_2}\right)$$

Consider a mass displaced through a vertical height Δh, where $\Delta h \ll$ radius of Earth R

$$\Delta E_{gp} = GMm\left(\frac{1}{R} - \frac{1}{(R + \Delta h)}\right)$$
$$\Delta E_{gp} = GMm\left(\frac{R + \Delta h - R}{R(R + \Delta h)}\right) = \frac{GMm\Delta h}{R^2} \quad R^2$$

$$\text{(because } R^2 \approx R^2 + R\Delta h\text{)}$$

$$\frac{GM}{R^2} = g, \text{ so } \Delta E_{gp} = mg\Delta h.$$

Worked example

A gas molecule high in the Earth's atmosphere travelling upward at the escape velocity will escape into space.

Calculate the kinetic energy that a hydrogen molecule must have in order to escape. (Radius of the Earth is 6400 km; mass of the Earth is 6.0×10^{24} kg; $G = 6.67 \times 10^{-11}\,\mathrm{N\,m^2\,kg^{-2}}$; molar mass of hydrogen = $0.002\,\mathrm{kg\,mol^{-1}}$; Avogadro constant $6.02 \times 10^{23}\,\mathrm{mol^{-1}}$). **(3 marks)**

$$v_{min} = \sqrt{\frac{2GM}{r}}$$

$$E_K = \frac{1}{2}mv^2 = \frac{0.5 \times 0.002}{6.02 \times 10^{23}} \times \frac{2GM}{r}$$

$$= 2.1 \times 10^{-19}\,\mathrm{J}$$

Now try this

1 Felix Baumgartner set a world record for the greatest vertical distance of free fall when he jumped to Earth from a helium balloon in the stratosphere in October 2012. The total distance he fell through in the jump was 39.0 km. Evaluate the accuracy of the approximate formula for gravitational potential energy changes, $\Delta E_{gp} = mg\Delta h$, in this situation. (Radius of the Earth = 6360 km; mass of the Earth is 5.98×10^{24} kg; $G = 6.67 \times 10^{-11}\,\mathrm{N\,m^2\,kg^{-2}}$; $g = 9.81\,\mathrm{N\,kg^{-1}}$.) **(6 marks)**

2 Satellites are usually moved into their final orbit in stages. Calculate the energy required to move a satellite of mass 5950 kg from an orbit at a height of 295 km to a geostationary orbit at a height of 35 800 km from the Earth's surface. **(3 marks)**

Exam skills

This exam-style question uses knowledge and skills you have already revised. Have a look at pages 105 and 106 for a reminder about orbits and gravitational fields.

Worked example

The Hubble Space Telescope orbits Earth in a circular orbit at an altitude of 550 km. The mass of the Hubble Space Telescope is 11 000 kg. The mass of the Earth is 6.0×10^{24} kg and the radius of the Earth is 6400 km. $G = 6.67 \times 10^{-11}$ N m^2 kg^{-2}.

(a) Suggest and explain one advantage of putting a telescope into orbit around the Earth. **(2 marks)**

The telescope can obtain clearer and more detailed images of astronomical objects because the light it receives does not have to pass through the Earth's atmosphere, which badly distorts light in optical wavelengths.

(b) Calculate the gravitational potential energy of the Hubble Space telescope. **(3 marks)**

$$GPE = -\frac{GMm}{r}$$
$$= \frac{-6.67 \times 10^{-11} \times 6.0 \times 10^{24} \times 11000}{(6.40 \times 10^6 + 0.55 \times 10^6)}$$
$$= -6.3 \times 10^{11} \text{ J}$$

(c) Calculate the time period of the orbit of the Hubble Space Telescope. **(4 marks)**

$$\frac{GMm}{r^2} = mr\omega^2 = \frac{4\pi^2 mr}{T^2}$$

$$T = \sqrt{\frac{4\pi^2 r^3}{GM}} = 5754 \text{ s} = 96 \text{ minutes.}$$

(d) (i) Calculate the total energy of the Hubble Space telescope in its orbit. **(4 marks)**

Total energy = gravitational potential energy + kinetic energy
$$E_K = \tfrac{1}{2}mv^2 = \tfrac{1}{2}m\left(\frac{2\pi r}{T}\right)^2$$
$$= \tfrac{1}{2} \times 11000 \times \left(\frac{2\pi \times 6.95 \times 10^6}{5754}\right)^2 = +3.2 \times 10^{11} \text{ J}$$
Total energy = $-6.3 \times 10^{11} + 3.2 \times 10^{11} = -3.1 \times 10^{11}$ J

(ii) Comment on the sign of the total energy and the relative sizes of the GPE and KE. **(2 marks)**

The total energy is negative because the satellite is trapped in an orbit – it does not have enough energy to escape from the Earth's gravitational field (to reach zero total energy).

(e) Although the Earth's atmosphere is extremely thin at the altitude of the Hubble Space Telescope, it does exert a small drag force on the satellite, and this force causes the orbit to decay. Discuss how this affects the total energy, gravitational potential energy, and kinetic energy of the satellite. **(3 marks)**

Total energy will fall because energy is transferred through heating. Gravitational potential energy will fall (become more negative) as the satellite gets closer to the Earth. Kinetic energy will increase, but not enough to compensate for the loss in gravitational potential energy.

Be careful! Altitude is measured from the surface of the Earth, but the distance needed when using the gravitational field equations must be measured from the centre of the Earth:
$r = R_E + h$
$= 6.40 \times 10^6 + 0.55 \times 10^6$
$= 6.95 \times 10^6$ m

'Suggest and explain' questions expect you to apply physics to an unfamiliar context. Always back up your suggestion with a valid explanation.

1 Quote the relevant equation.
2 Substitute values into the equation.
3 Carry out the calculation and give the answer.
4 Include correct S.I. units.
5 Watch out for the sign!

Show each stage of a complex calculation, using algebra to highlight the key steps.

This question asks you to comment on the sign, so go back and check that you have not made a sign error or lost a minus sign in the earlier parts.

The gravitational potential energy and total energy must fall for the reasons stated, but what about kinetic energy?

Remember that equation for orbital speed is $\left(v = \sqrt{\frac{GM}{r}}\right)$ and this is greater for smaller orbits, so kinetic energy is greater.

Formation of stars

Stars form when gravitational forces cause dust and gas clouds to collapse into a single mass.

Billions of suns

The Earth is one of eight planets that orbit the Sun, our nearest star.

The Sun is one of about 300 billion stars in our galaxy, up to a third of which may have their own solar systems.

There may be as many as 100 billion galaxies in the observable Universe, so our Solar System is far from unique in the Universe.

In addition to the eight planets, up to 50 dwarf planets (such as Pluto) and millions of other minor planets and comets orbit the Sun. (Diagram is not to scale.)

Star formation

Clouds of dust and gas are scattered throughout space. Turbulence within these clouds produces accretions of sufficient mass that the gas and dust can begin to collapse under their own gravitational forces.

As the cloud collapses the gravitational forces do work. There is an energy transfer from the gravitational field to internal energy of the dust cloud, and so the cloud begins to heat up.

As the cloud continues to collapse, a dense, hot core forms. It is this hot core at the heart of the collapsing cloud that may become a star.

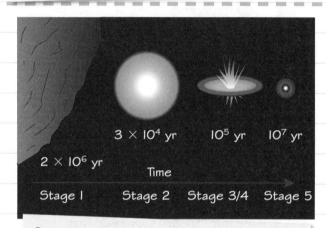

3×10^4 yr 10^5 yr 10^7 yr

2×10^6 yr

Time

Stage 1   Stage 2   Stage 3/4   Stage 5

Some of the dust and gas in a collapsing cloud can become planets, planetary satellites (moons), comets, or asteroids.

Worked example

Calculate the gravitational potential energy transferred if Jupiter (mass 1.90×10^{27} kg) were to be brought to the surface of the Sun. The radius of Jupiter's orbit about the Sun is 7.80×10^8 km. ($G = 6.67 \times 10^{-11}$ N m^2 kg^{-2}; $M_S = 1.99 \times 10^{30}$ kg; $r_S = 7.0 \times 10^5$ km) **(3 marks)**

$\Delta E_{gp} = GMm\left(\dfrac{1}{r_1} - \dfrac{1}{r_2}\right)$

$= 6.67 \times 10^{-11} \times 1.99 \times 10^{30} \times 1.90 \times 10^{27}$

$\times \left(\dfrac{1}{7.8 \times 10^{11}} - \dfrac{1}{7.0 \times 10^8}\right)$

$= 3.6 \times 10^{38}$ J

Planetary satellites

The Earth has a single moon, but some other planets have many more. Jupiter, the largest planet in the Solar System, has more than 60 moons, but Venus has none.

Maths skills Take care with powers of 10. In this example distances are given in km, but they must be converted to m (by multiplying the values by 10^3) before substituting into the equation.

Now try this

1 Jupiter is about 300 times more massive than the Earth. A student suggests that on the surface of Jupiter an object would weigh about 300 times more than it would weigh on the surface of the Earth. However, a textbook states that on the surface of Jupiter an object would weigh about 3 times more than it would weigh on the Earth. Write an explanation to help the student understand how this can be. **(2 marks)**

2 The Moon is moving away from the Earth at a rate of a few millimetres per year. Calculate the gravitational potential energy that will be transferred over the next couple of centuries as the Moon moves 10 m further away from the Earth. ($M_E = 5.97 \times 10^{24}$ kg, $m_M = 7.35 \times 10^{22}$ kg; the (mean) radius of the Moon's orbit about the Earth is 3.84×10^5 km; $G = 6.67 \times 10^{-11}$ N m^2 kg^{-2}.) **(3 marks)**

Evolution of stars

A star's life cycle depends upon its mass.

Fusion

The hot core at the centre of the collapsing cloud is known as a **protostar**. As the protostar accumulates gas and dust, its density and temperature increase. Once the core reaches a temperature of about 10^7 K, hydrogen nuclei begin to fuse into helium nuclei, releasing large amounts of energy.

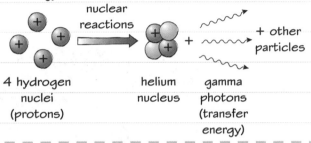

4 hydrogen nuclei (protons)   nuclear reactions   helium nucleus   gamma photons (transfer energy)   + other particles

Pressure inside stars

The gas making up a star exerts **gas pressure**. Once hydrogen fusion has begun, the photons released exert a pressure called **radiation pressure**. Eventually this outward pressure is large enough to halt the gravitational collapse of the cloud.

The protostar has become a stable **main-sequence star**. It will remain in this state for millions or even billions of years, depending upon its mass.

gravitational force pulling inwards

gas and radiation pressure pushing

Working scientifically

Before the process of nuclear fusion was understood, people were puzzled about how the Sun and stars radiated energy.
A 19th century theory supposed that the Sun was made of burning carbon; this would have given the Sun an age of less than 5000 years. Lord Kelvin proposed alternative explanations and concluded that the Sun could shine for up to 45 million years. However, the fossil record implies that the Earth has to be at least 300 million years old. Only in the 20th century was a satisfactory explanation for the energy source of the Sun developed.

Running out of hydrogen

When most of the hydrogen in the core of a low to medium mass star (up to four times the Sun's mass) is used up, the radiation pressure decreases. The star contracts under gravitational forces and hydrogen-to-helium fusion begins in the star's outer layers.

The energy released in the helium shell causes radiation pressure that makes the outer layer of the star expand. As the surface area of the star increases, its surface temperature decreases. The larger, cooler star is known as a **red giant**.

The helium flash

Meanwhile, the core continues to collapse and its temperature rises. Once the temperature reaches 10^8 K, a helium flash occurs: helium in the core fuses rapidly to give carbon and oxygen. Most of the material around the core is ejected as a glowing cloud of gas called a **planetary nebula**.

The remnant of the core glows brightly at first, but no further fusion takes place. This high-density object is known as a **white dwarf**. The white dwarf cools down, and is predicted to become a black dwarf after many billions of years.

Worked example

For nuclear fusion to occur, protons must have sufficient kinetic energy to overcome their electrostatic repulsion, and hence must have a very high temperature. Estimate the temperature required to initiate fusion by calculating the energy required to bring two protons to a separation of 10^{-15} m.
($\varepsilon_0 = 8.85 \times 10^{-12}$ C^2 N^{-1} m^{-2}; $e = 1.60 \times 10^{-19}$ C; $k = 1.38 \times 10^{-23}$ J K^{-1})   **(4 marks)**

$$W = \frac{Q_1 Q_2}{4\pi\varepsilon_0 r}$$

$$= \frac{(1.6 \times 10^{-19})^2}{4\pi \times 8.85 \times 10^{-12} \times 1 \times 10^{-15} \, m}$$

$$= 2.3 \times 10^{-13} \, J$$

$$E_k = \tfrac{3}{2}kT = 2.3 \times 10^{-13} \, J$$

$$\therefore T = \frac{2 \times 2.3 \times 10^{-13}}{3 \times 1.38 \times 10^{-23}} = 1.1 \times 10^8 \, K$$

For this question you will need to use the equation for work done in an electric field due to a point charge. $W = \frac{Qq}{4\pi\varepsilon_0 r}$, covered on page 130.

Now try this

Draw a flow chart to show the evolutionary path of the Sun from protostar through to its end point.
(2 marks)

End points of stars

The end point of a star depends upon its mass while it is on the main sequence.

White dwarfs

White dwarfs, remnants of cores of low-to-medium mass stars, have approximately the mass of the Sun but the radius of Earth. This high density means that the gravitational field strength on the surface of a white dwarf is about 350 000 times that at the surface of the Earth.

Fusion does not take place in a white dwarf, but the collapse of the core remnant is halted by quantum mechanical repulsion between electrons in the remnant, called degeneracy pressure.

Massive core remnants

As a massive star reaches the end of its time on the main sequence it becomes a **red supergiant**, rather than a red giant.

When the core of a supergiant collapses, gravitational forces overcome the degeneracy pressure as the core mass exceeds the Chandrasekhar limit. Electrons within the core combine with protons to produce neutrons and neutrinos. The neutrinos escape, leaving an inner core filled with neutrons.

The outer shells rapidly collapse and rebound against the neutron core. The shock wave generated expels the surface layers in a violent explosion known as a **supernova**.

The Crab nebula is a supernova remnant.

Degeneracy pressure

If electrons are close enough together, they are forced into their lowest possible energy levels until all the available levels are filled and it is not possible to add more electrons. This is called a **degenerate** state. The repulsion between the electrons causes a **degeneracy pressure**, preventing the core remnant from collapsing.

The degeneracy pressure can only maintain the size of the core remnant up to 1.4 solar masses. This is known as the **Chandrasekhar limit**. For more massive core remnants the weight of the star overcomes the electron degeneracy pressure, and the collapse continues.

Neutron stars and black holes

The remnant of a supernova is a **neutron star**. Incredibly dense and spinning at a fast rate, the neutron star emits beams of electromagnetic radiation. If the magnetic poles align with the Earth, we detect a pulse of radiation on each revolution. Such neutron stars were the first to be discovered and are called **pulsars**.

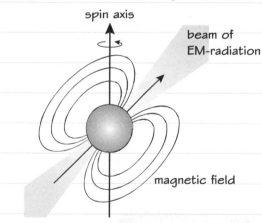

The most massive supergiants have cores that continue to collapse after the supernova explosion. Eventually the escape velocity exceeds the speed of light, and the star has become a **black hole** from which nothing can escape, not even radiation.

Worked example

The Schwarzschild radius is the radius of a sphere such that, if all the mass of an object were within the sphere, the escape velocity from it would be the speed of light.

(a) Derive an expression for this radius in terms of the mass of the object. **(3 marks)**

$$\Delta E_K = -\Delta E_{gp} \quad \therefore \frac{1}{2}mv^2 = \frac{GMm}{r}$$

$$\therefore r = \frac{2GM}{v^2}$$

At the Schwarzschild radius $v = c$ $\therefore r = \frac{2GM}{c^2}$

(b) Calculate the Schwarzschild radius for an object of 3 solar masses. ($G = 6.67 \times 10^{-11}\,\mathrm{N\,m^2\,kg^{-2}}$; $M_S = 1.99 \times 10^{30}\,\mathrm{kg}$; $c = 3.00 \times 10^8\,\mathrm{m\,s^{-1}}$.) **(2 marks)**

$$\therefore r = \frac{2GM}{c^2} = \frac{2 \times 6.67 \times 10^{-11} \times 3 \times 1.99 \times 10^{30}}{(3.0 \times 10^8)^2}$$

$$\therefore r = 8.9 \times 10^3\,\mathrm{m}$$

Now try this

Compare and contrast a planetary nebula and a supernova remnant. **(4 marks)**

The Hertzsprung-Russell diagram

The Hertzsprung–Russell (HR) diagram is a plot of star luminosity against surface temperature that shows several groups of stars.

The Hertzsprung-Russell diagram

The HR diagram is a scatter graph plotting luminosity (the rate of radiation of light) against surface temperature for stars.

The large range of luminosities means that a logarithmic scale is used for the luminosity axis.

The temperature range is smaller, but the scale is usually logarithmic too. Originally the temperature scale was a scale based on the spectral class of the stars, and so the temperature scale runs in the reverse direction.

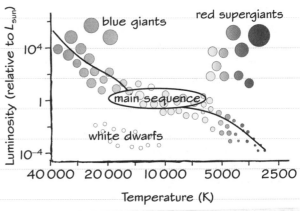

You should know approximate values for temperature and luminosity values on the HR diagram.
L_{sun} = luminosity of our Sun.

Main-sequence stars

In the two previous topics we looked at stars in the **main sequence**, – stars that are stable, fusing hydrogen nuclei into helium nuclei in their cores. In the HR diagram this main sequence is shown, with star mass increasing from bottom right to top left. The more massive the star, the shorter the time it spends on the main sequence.

Stars of average mass (such as our Sun) will remain on the main sequence for up to 10 billion years. The massive blue giant stars found at the top left of the main sequence may only spend a few million years on the main sequence, as they exhaust their supply of hydrogen in the core much more quickly than less massive stars.

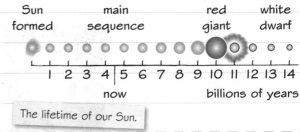

The lifetime of our Sun.

Red giants and white dwarfs

The red giants and red supergiants are located at the top right of the HR diagram. These are cool stars with a very large surface area. Hence they have a high luminosity despite their low surface temperature.

White dwarfs are located towards the bottom left of the HR diagram. Although they have a high surface temperature, their surface area is relatively small. Hence they have a low luminosity despite their high surface temperature.

Use the HR diagram above to calculate the relative difference in luminosity between a low-mass and a high-mass main-sequence star. **(2 marks)**

Low-mass star has a luminosity $\approx 10^{-4} \times L_{sun}$

High-mass star has a luminosity $\approx 10^{4} \times L_{sun}$

So relative difference $\approx \dfrac{10^{-4} \times L_{sun}}{10^{4} \times L_{sun}} = 10^{8}$

Maths skills

$0.001 = 10^{-3}$	$\log_{10}(0.001) = -3$
$1 = 10^{0}$	$\log_{10}(1) = 0$
$1000 = 10^{3}$	$\log_{10}(1000) = 3$
$1\,000\,000 = 10^{6}$	$\log_{10}(1\,000\,000) = 6$

'$\log_{10}$' is often simply written 'log'.

Do not confuse this with $\log_{e}$ or ln, which represents natural (base e) logs.

Massive stars such as blue giants only spend a comparatively short time as main-sequence stars. Explain why such large mass stars burn the hydrogen in their core more quickly than smaller mass main-sequence stars. **(2 marks)**

Energy levels in atoms

Electrons in atoms can have only certain discrete amounts of energy – the energy is quantised.

Energy states

Following the emergence of a nuclear model of the atom, Bohr proposed a model that explained how electrons could have stable orbits around the nucleus. He suggested that the angular momentum of the electrons was **quantised**: the electrons moved only in orbits of specific radii with defined energy.

The energy of the electron depends on the size of the orbit. All energy states are negative, since the electron is attracted to the nucleus and energy must be supplied to remove the electron from the atom.

The ground state

The lowest energy state ($n = 1$) is referred to as the ground state. Higher energy states are called excited states.

In a stable atom all electrons are in their lowest possible energy states.

Single atoms have simple energy level structures.

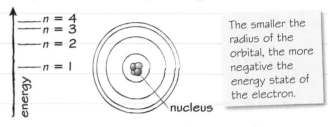

The smaller the radius of the orbital, the more negative the energy state of the electron.

Changing between energy states

If energy is supplied to the atom, electrons can be promoted to higher energy states. These higher energy states are unstable, and the electrons will return to their original energy states.

Radiation is emitted when the electron falls from a higher to a lower energy state. The difference between the energy states, ΔE, determines the frequency f of the radiation emitted.

$\Delta E = hf$, or $\Delta E = \dfrac{hc}{\lambda}$, where h is the Planck constant, 6.63×10^{-34} J s

If a gas is heated the spectrum produced reveals its energy level structure. Each element has its own energy level structure, so emission spectra are like fingerprints.

Hydrogen emission spectrum

Each wavelength emitted in the hydrogen **emission line spectrum** here is produced by electrons undergoing a transition to the $n = 2$ energy levels.

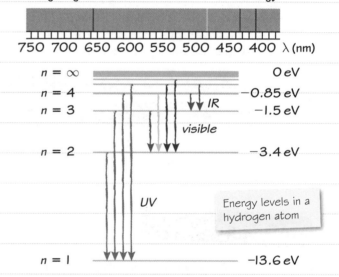

Energy levels in a hydrogen atom

Worked example

Calculate the wavelength of light emitted when an electron in a hydrogen atom drops to the ground state from the first excited state. **(3 marks)**

$\Delta E = \dfrac{hc}{\lambda}$

$\therefore \lambda = \dfrac{6.63 \times 10^{-34} \times 3.00 \times 10^{8}}{(-3.40 - (-13.60)) \times 1.6 \times 10^{-19}}$

$\therefore \lambda = 1.22 \times 10^{-7}$ m

$= 122$ nm

Worked example

Calculate the energy in joules required to ionise a hydrogen atom. (1 eV $= 1.6 \times 10^{-19}$ J.) **(2 marks)**

$\Delta E = 13.6\,eV = 13.6 \times 1.6 \times 10^{-19}$ J

$= 2.2 \times 10^{-18}$ J

Now try this

Rutherford's model of the atom consisted of electrons orbiting a nucleus. This model was not able to explain atomic spectra. The Bohr model was more successful in this respect. Explain how the Bohr model was able to explain atomic spectra, whereas the Rutherford model could not. **(2 marks)**

Emission and absorption spectra

The light from a star can give important information about what the star is made of.

Emission spectra

When low-pressure monatomic gases are heated, a series of discrete wavelengths is emitted. This is an **emission line spectrum**.

When polyatomic gases and liquids are heated, a band emission spectrum is produced. Band spectra consist of more wavelengths than are seen in a line spectrum, linked in bands.

Hot dense solids emit a **continuous spectrum** with wavelengths across a wide range. The range of the spectrum relates to the temperature of the hot solid.

A monatomic gas has a clear emission spectrum of lines.

neon

A diatomic gas has an emission spectrum of bands.

nitrogen

hot solid

4000   5000   6000   7000 $\lambda / \times 10^{-10}$ m

Line, band and continuous emission spectra, observed by passing light through a transmission diffraction grating.

Absorption spectra

If an atom emits particular wavelengths when excited electrons drop back into more stable energy levels, the atom will also absorb light of these wavelengths when electrons are promoted to higher energy levels.

If light of a continuous range of wavelengths is passed through a gas or liquid, a series of missing wavelength lines or bands will be observed. The missing wavelengths have been absorbed by the material the light has passed through.

An emission line spectrum and the matching absorption line spectrum

Identifying elements in stars

The light emitted from a star is a continuous spectrum. However, elements in the outer region of the star absorb light of particular wavelengths so an absorption line spectrum is produced.

Hence the elements present in the star can be determined, even if it is hundreds of light-years away. This tells us the type of star being observed.

The element helium was first discovered from the dark lines it left in the Sun's continuous spectrum.

Worked example

Light from a hot solid is passed through a low-pressure gas and a series of dark lines is seen to cross the continuous spectrum from the hot solid. Explain how these dark lines are produced. **(6 marks)**

Some of the light from the hot solid has the same wavelength (and therefore energy) as the light that would be emitted by excited atoms of the low-pressure gas.

These photons are absorbed by the low-pressure gas, causing electrons to be promoted to excited energy states.

When the electrons drop back down to their ground state, energy is radiated away. This light is re-radiated in all directions equally. Therefore in the original direction the light intensity is reduced for these wavelengths.

Hence a series of dark lines crosses the continuous spectrum.

Worked example

Use the spectra provided to identify some of the gases present in the nebula being observed. **(2 marks)**

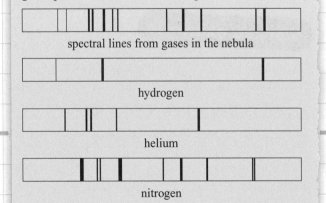

spectral lines from gases in the nebula

hydrogen

helium

nitrogen

The nebula contains H and He, since the lines in the spectra of these two elements are present in the spectrum of light from the nebula.

Now try this

Devise an experiment to investigate the emission spectrum of an element using a low-pressure gas discharge tube and a diffraction grating. Include a labelled diagram, and a description of how you would process your results. **(6 marks)**

Wien's law and Stefan's law

The spectrum emitted by a star enables its surface temperature and diameter to be determined.

Black-body radiation

A **black body** is one that absorbs all the EM radiation that is incident upon it. To stay in thermal equilibrium, it must emit radiation at the same rate as it absorbs it. A black body is therefore a perfect radiator.

The radiation emitted from a black body depends on the surface temperature of the black body, T. The spectrum has a characteristic shape, with a peak power emission at a particular wavelength, λ_{max}.

The relationship between the wavelength at which peak power radiation occurs and the temperature is known as Wien's displacement law:

$L = A\sigma T^4$

$\lambda_{max} \propto \dfrac{1}{T}$

$\therefore \lambda_{max}T = 2.9 \times 10^{-3}\,\text{m K}$

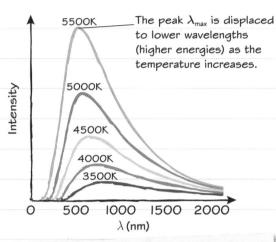

The peak λ_{max} is displaced to lower wavelengths (higher energies) as the temperature increases.

Emission from a black body at various temperatures

Stefan's law

The power radiated from a black body is a function of its temperature. Specifically, Stefan's law states that the total power radiated across all wavelengths per unit surface area of a black body is proportional to the fourth power of the body's temperature T.

Stars are approximate black body emitters, and so we can relate their luminosity (power) L to their area

$A = 4\pi r^2$

$L = 4\pi r^2 \sigma T^4$

where $\sigma = 5.67 \times 10^{-8}\,\text{W m}^{-2}\,\text{K}^{-4}$.

Worked example

Most stars are too far away for us to be able to measure their diameter by observation.
How do we know that red giant stars are much bigger than the main-sequence stars from which they form? **(4 marks)**

When a star moves off the main sequence, λ_{max} shifts to the red end of the spectrum, indicating that the surface of the star has cooled. At the same time, its luminosity L increases enormously.

$L = A\sigma T^4$ so if L has increased whilst T has decreased, then the area A must have increased. Hence the star has become much bigger.

Worked example

A star has a surface temperature of 5800 K.

(a) Calculate the wavelength at which peak power radiation occurs. **(2 marks)**

$\lambda_{max}T = 2.9 \times 10^{-3}\,\text{m K}$

$\lambda_{max} = \dfrac{2.9 \times 10^{-3}}{5800} = 5.0 \times 10^{-7}\,\text{m}$

(b) Compare the luminosity of this star with one of similar size, but a temperature of 20 000 K. **(2 marks)**

$\dfrac{L_1}{L_2} = \dfrac{T_1^4}{T_2^4} = \left(\dfrac{5800}{20\,000}\right)^4 = 0.0071$

The first star has <1% of the luminosity of the hotter star.

Now try this

1. Use Stefan's law to calculate the surface area of the Sun. (The Sun's power output is 3.6×10^{26} W and its surface temperature is 5800 K.) **(2 marks)**

2. The star Alioth, or ε Ursae Majoris, is the brightest star in the constellation of the Plough. It has a peak spectral emission at $\lambda_{max} = 269$ nm and a luminosity of 3.89×10^{28} W. Calculate its surface temperature and radius. ($\sigma = 5.67 \times 10^{-8}\,\text{W m}^{-2}\,\text{K}^{-4}$.) **(6 marks)**

The distances to stars

For the huge distances in astronomy, we use alternative distance units to the metre and indirect measuring methods.

Astronomical units

Distances to objects in our Solar System are very large. For example, the distance from the Sun to Jupiter is about 8×10^{11} m. It can be simpler to express this in terms of the distance from the Sun to the Earth, i.e., in **astronomical units** ($1 \, AU = 1.50 \times 10^{11}$ m). Jupiter is just over 5 AU from the Sun.

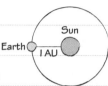

One astronomical unit (AU) is the distance from the Earth to the Sun.

Parallax

As the Earth revolves around the Sun, nearby stars appear to move against the background of more distant stars. This is referred to as annual **parallax**, and can be used to determine the distances to nearby stars.

Astronomers make two measurements of a nearby star's position 6 months apart. The change in angular position allows the distance to the star to be calculated using trigonometry.

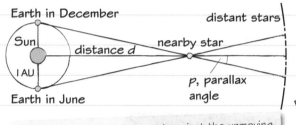

Earth in December

Sun

1 AU

distance d

nearby star

distant stars

p, parallax angle

Earth in June

The angle p can be measured against the unmoving background of distant stars, and the distance from the Earth to the Sun is known.

Parsecs and light-years

One angular degree is split into 60 minutes of arc, and one minute is split into 60 seconds of arc, or arcseconds.

Astronomers often use the reciprocal of the parallax angle p as the distance. If p is expressed in seconds of arc, then the distance d to the star in **parsecs** (pc) is:

$$d = \frac{1}{p}$$

$1 \, pc = 3.09 \times 10^{16}$ m

Another unit favoured by astronomers is the **light-year** (ly). This is simply the distance travelled by light in a vacuum in a year.

$1 \, ly = 9.46 \times 10^{15}$ m

The farther away a star is, the smaller its parallax angle. Hence parallax is only suitable for nearby stars – up to about 125 pc.

Worked example

From basic principles, show that 1 pc is equal to about 3 ly. (1 AU is 1.50×10^{11} m.) **(4 marks)**

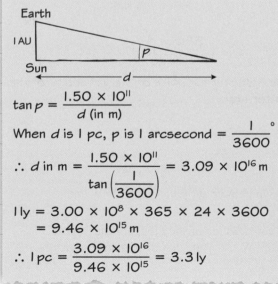

Earth

1 AU

p

Sun

d

$$\tan p = \frac{1.50 \times 10^{11}}{d \text{ (in m)}}$$

When d is 1 pc, p is 1 arcsecond $= \dfrac{1}{3600}°$

$$\therefore d \text{ in m} = \frac{1.50 \times 10^{11}}{\tan\left(\frac{1}{3600}\right)} = 3.09 \times 10^{16} \text{ m}$$

$1 \, ly = 3.00 \times 10^8 \times 365 \times 24 \times 3600$

$\qquad = 9.46 \times 10^{15}$ m

$$\therefore 1 \, pc = \frac{3.09 \times 10^{16}}{9.46 \times 10^{15}} = 3.3 \, ly$$

Maths skills In the equation $d = \dfrac{1}{p}$ the angle is in arcseconds.

1 arcsecond $= \dfrac{1}{60 \times 60} = 2.78 \times 10^{-4}$ degrees

For such small angles in radians, $\sin\theta \approx \tan\theta \approx \theta$

This **small-angle approximation** is often useful, but it is only valid if the angle is expressed in radians. So to use the approximation in this example we would have to express 1 arcsecond in radians.

$$\text{angle in radians} = \frac{\text{angle in degrees}}{360} \times 2\pi$$

Now try this

Explain why annual parallax is only suitable as a method of determining distances to the closest stars. **(2 marks)**

The Doppler effect

Whenever there is relative movement between a source of waves and an observer, a change in the wavelength (and frequency) of the waves is observed.

The Doppler effect

When the wave source and the observer are moving closer to each other, the observed wavelength decreases and the frequency increases.

When the source and observer are moving apart, the observed wavelength increases and the frequency decreases.

The effect is observed with all types of waves and has important applications in medical diagnosis with ultrasound, and in astronomy.

waves propagating from a stationary source

Relative motion of a wave source changes the wavelength noted by the observer.

waves propagating from a moving source

receding source: the wavelength appears to be bigger.

approaching source: the wavelength appears to be smaller.

relative motion ➡

Doppler shift for light

The change in the observed wavelength, the **Doppler shift**, depends on the speed of the waves in the medium as well as the relative speed of the observer and source in the medium.

For EM radiation, the speed of the waves is very large compared with the speed of the source/observer, so we can use the approximate formula

$$\frac{\Delta\lambda}{\lambda} \approx \frac{\Delta f}{f} \approx \frac{v}{c}$$

where c is the speed of light and v is the speed of the source relative to the observer.

λ (or f) is the wavelength (or frequency) you would measure for light from a similar source in the laboratory.

Doppler shift in astronomy

The light from many astronomical objects is Doppler shifted.

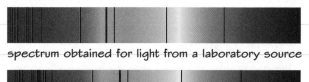

spectrum obtained for light from a laboratory source

spectrum obtained for light from a galaxy

The absorption line spectrum from the galaxy has been Doppler shifted.

Absorption line spectra are explained on page 114.

In the lower absorption spectrum, the wavelengths of the dark lines have been shifted. The magnitude of the shift depends on the speed of the source:

$$\Delta\lambda \approx \left(\frac{v}{c}\right) \times \lambda$$

Worked example

A particular line in the spectrum of light from the Triangulum galaxy shows a fractional decrease in wavelength of -6.1×10^{-4}, and light from the Black Eye galaxy has a fractional increase of $+1.4 \times 10^{-3}$.

State which galaxy is moving faster, and which one is moving away from us. **(4 marks)**

The Black Eye galaxy is moving faster, as the fractional change is greater and $\Delta\lambda \propto v$. It is also moving away from us, since the wavelength is increased.

Worked example

Explain two things the Doppler shift can tell us about an object's relative motion. **(3 marks)**

The Doppler shift tells us the relative speed of the object, as $\Delta\lambda \propto v$.

It also tells us whether the source is approaching or receding from the observer, as the wavelength increases for recession and decreases for approach.

Now try this

A motorist is caught driving through a red traffic light at a junction. He claims that the red light was Doppler shifted so that he observed the red light as green. Calculate how fast he would have to be travelling for this statement to be correct ($\lambda_{red} = 635$ nm; $\lambda_{green} = 565$ nm). **(4 marks)**

Hubble's law

Hubble's law provides evidence for an expanding Universe and thus the Big Bang theory.

Redshift

Vesto Slipher was the first person to observe the shift in spectral lines of galaxies, which at the time were thought to be nebulae within our own galaxy. Most galaxies observed had **redshifts** – the lines shifted to longer wavelengths.

Hubble determined the distances to these galaxies. He found an approximate proportionality between Slipher's measurements of the galaxies' redshifts $z = \dfrac{\Delta\lambda}{\lambda}$ and their distances d from the Earth, $z \propto d$.

The observation that every distant galaxy in the Universe is redshifted tells us that all galaxies beyond our local area are moving away from us.

This conclusion relies on the **cosmological principle** – the assumption that the Universe is homogeneous and isotropic (the same in all directions) and the laws of physics are universal.

Hubble's law

Hubble concluded that all galaxies were receding from the Earth according to the expression

$v \approx H_0 d$, where H_0 is the Hubble constant.

This is Hubble's law, which tells us that the further away a galaxy is, the faster it is receding. This was the first evidence of an expanding Universe.

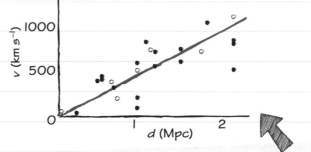

Each dot represents a galaxy. They do not all fall onto the line $v = H_0 d$ because their relative motion away from us is combined with motion in other directions.

The Big Bang

Since galaxies are moving away from each other, we can imagine 'running the film backwards'. This would show galaxies all converging at one point. This origin of the Universe is thought to have taken place approximately 13.8 billion years ago, when the Universe expanded from an infinitely high-density state in a so-called **Big Bang**.

The Universe has been expanding ever since, with gravity acting as a brake to slow down the expansion rate.

Alongside Hubble's law, another key piece of supporting evidence for the Big Bang is the **cosmic microwave background** (CMB). This is radiation from soon after the Big Bang. Ever since the Big Bang, the space–time of the Universe itself – not merely the matter in it – has been expanding, thus stretching the wavelength of the CMB radiation. Its current wavelength is characteristic of black-body radiation from an object at a temperature of 2.7 K.

The value of the Hubble constant is not known with certainty but is thought to be around 67–$73\,\text{km s}^{-1}\,\text{Mpc}^{-1}$. The Hubble constant can also be expressed in s^{-1}, if v and d are expressed with the same units of distance.

The expanding Universe

The fact that we see all distant galaxies moving away from us does not mean that the Earth is at the centre of the Universe. In an expanding Universe all galaxies are receding from other galaxies.

A rising currant bun is a good model:

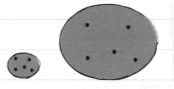

Each currant moves away from all other currants as the bun expands, because the dough itself is getting bigger.

Worked example

The Pinwheel galaxy is 6.4 Mpc from the Earth and has a redshift of 1.71×10^{-3}.

Use this information to calculate a value for the Hubble constant in $\text{km s}^{-1}\,\text{Mpc}^{-1}$.   **(2 marks)**

$z = \dfrac{\Delta\lambda}{\lambda} = \dfrac{v}{c}$

$v = zc = 1.71 \times 10^{-3} \times 3.00 \times 10^8$
$\quad\; = 5.13 \times 10^5\,\text{m s}^{-1}$

$v \approx H_0 d$

$\therefore H_0 = \dfrac{5.13 \times 10^2}{6.4} = 80\,\text{km s}^{-1}\,\text{Mpc}^{-1}$

Now try this

Suggest why the value for H_0 calculated in the example above may be inaccurate.   **(4 marks)**

The evolution of the Universe

The ultimate fate of the Universe depends on the amount of dark energy in the Universe.

The start of the Universe

Immediately after the Big Bang, the Universe was dense, hot (> 10^{22} K), and almost infinitely small. In a tiny fraction of a second, space–time expanded (**inflation**); then particles and antiparticles were formed – but with very slightly more particles, so that as they collided and annihilated, a Universe made from particles remained. The annihilation produced high-energy photons that were repeatedly absorbed and re-emitted by the charged particles.

Still in less than a second, the Universe cooled enough for the particles to form protons and neutrons. After a minute, the temperature had dropped so much that no further fusion occurred, but it took another 250 000 years for the temperature to fall enough for electrons and nuclei to combine into hydrogen and helium atoms. Photons could now travel freely through space, since neutral atoms do not interact so readily with photons. These photons became the CMB radiation.

After 1 million years, stars and galaxies began to form. After a billion years, the gravitational collapse of these first stars formed heavy elements.

Dark matter and dark energy

Newton's laws predict that the stars closest to the centre of a galaxy should move faster than those out on the edge, but measurements indicate that all stars orbit the centres of galaxies at about the same speed. This suggests the presence of matter that exerts a gravitational pull on all 'normal' matter, but does not emit EM radiation and so cannot be 'seen' – hence '**dark' matter**.

Recently astronomers have detected an acceleration in the expansion rate of the Universe, rather than the deceleration expected due to gravity. They attribute this to **dark energy**, which, they hypothesise, introduces a repulsive force that shows itself only on the largest cosmic scale. With dark energy present, the expansion rate of the Universe initially slows because of the effect of gravity, but then increases. The ultimate fate of the Universe is then that it will expand forever.

At present neither dark matter nor dark energy is understood.

Worked example

Show that the reciprocal of the Hubble constant is equal to the time elapsed since the Big Bang, and hence estimate the age of the Universe.
Use $H_0 = 71 \text{ km s}^{-1} \text{ Mpc}^{-1}$. (1 pc $\approx 3.1 \times 10^{16}$ m.)

(4 marks)

For a galaxy with speed $v = H_0 d = \dfrac{d}{t}$

$\therefore \dfrac{1}{H_0} = t$

$1 \text{ Mpc} = 3.1 \times 10^{22} \text{ m}$

$\therefore H_0 = \dfrac{71 \times 10^3}{3.1 \times 10^{22}} \text{ s}^{-1}$

$\therefore t = \dfrac{1}{H_0} = \dfrac{3.1 \times 10^{22}}{71 \times 10^3} = 4.37 \times 10^{17} \text{ s}$

$= 13.8 \times 10^9 \text{ years}$

Matter–energy proportions

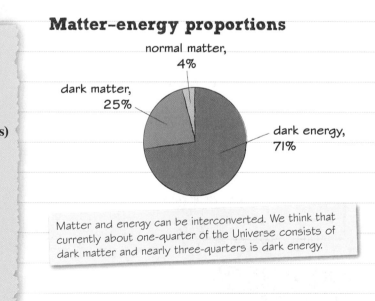

normal matter, 4%

dark matter, 25%

dark energy, 71%

Matter and energy can be interconverted. We think that currently about one-quarter of the Universe consists of dark matter and nearly three-quarters is dark energy.

Now try this

Suggest why the value for the age of the Universe calculated in the example above may be inaccurate.   **(4 marks)**

Capacitors

Charging a capacitor temporarily stores energy for later use.

Structure of a capacitor

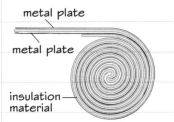

metal plate

metal plate

insulation material

A capacitor stores energy by separating charges onto two electrical conductors separated by an insulator.

Capacitors are used in back-up energy supplies, for example in data centres, or where fast, large p.d. discharges are needed, such as camera flashes or defibrillators.

Energy stored in a capacitor

Work is done as charge moves through the net p.d. in the circuit. Thus energy is stored in the electric field between the plates of the capacitor.

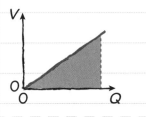

The work done is equal to the shaded area under the graph of p.d. between the plates, V, against charge on the plates, Q.

Charging a capacitor

When the switch is closed, electrons flow onto one plate of the capacitor and move off the other plate. The p.d. across the plates increases until it is equal to the p.d. across the cells. The flow of electrons then stops and the capacitor is fully charged. When the source of e.m.f. is removed, electrons flow back from the negatively charged plate of the capacitor around the circuit to the positively charged plate until the p.d. across the capacitor is zero.

The amount of charge stored by the capacitor is directly proportional to the p.d. between the plates. The amount of charge, Q, that can be stored per unit p.d., V, is called the **capacitance**, C (units farad, F).

$$C = \frac{Q}{V}$$

A 1 F capacitor carries a charge of 1 C when it has a p.d. of 1 V.

🖩 Maths skills | Calculating stored energy

$W = QV_{av}$, where V_{av} is the *average* p.d. through which the charge moves.

Since $Q \propto V$, the average p.d. is $\frac{V}{2}$ where V here is the final p.d. across the charged capacitor.

So the work done is given by $W = Q \times \frac{1}{2}V$, the area under the graph.

Hence the energy stored can be calculated using

$$W = \frac{1}{2}QV$$

Since $Q = CV$ we can also write

$$W = \frac{1}{2}CV^2 \text{ and } W = \frac{1}{2}\frac{Q^2}{C}.$$

Worked example

The dome of a van de Graaff generator has a capacitance of 15 pF. Calculate the energy stored when the dome is raised to a potential of 150 kV. **(2 marks)**

$$W = \frac{1}{2}CV^2 = 0.5 \times 15 \times 10^{-12} \times (150 \times 10^3)^2$$
$$= 0.17 \text{ J}$$

🖩 **Maths skills** Remember to convert values into S.I. units before substituting into an equation.

Now try this

When someone's heart stops beating effectively, a defibrillator can be used to deliver an electrical current through the chest, which is intended to shock the heart back into a normal rhythm. The energy for the electric current is stored in a capacitor. The capacitor in a certain defibrillator has a capacitance of 1200 μF, and the energy stored when the capacitor is fully charged is 125 J.

(a) Show that the potential difference needed to store this amount of energy in the capacitor is about 500 V. **(2 marks)**

(b) Calculate the initial discharge current if the discharging circuit has a total resistance of 150 Ω. **(2 marks)**

Series and parallel capacitor combinations

Combining capacitors in different arrangements changes the overall capacitance.

Parallel combination

When the switch is closed, charge flows until the p.d. across the capacitors exactly opposes the e.m.f. of the cells.

Once charging is complete the p.d. across each capacitor is V.

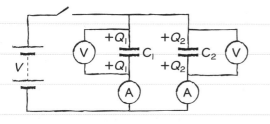

Calculation for capacitors in parallel

The charge on capacitors C_1 and C_2 is given by $Q_1 = C_1V$ and $Q_2 = C_2V$, respectively.

The total charge is given by $Q = Q_1 + Q_2$

So $Q = C_1V + C_2V$

Hence the capacitance of the combination is given by $C = \dfrac{Q}{V} = \dfrac{C_1V + C_2V}{V}$

$C = C_1 + C_2$

Series combination

When the switch is closed, charge flows until the net p.d. is 0 V (the p.d. across the capacitors exactly opposes the e.m.f. of the cells).

Since p.d.s add in a series circuit, we can write:

$V = V_1 + V_2$

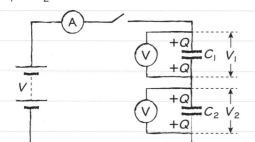

Since the capacitors are in series, the current in each part of the circuit is the same.

Hence at each stage of the charging process each capacitor has the same charge.

Approach questions with capacitor combinations in a similar way to questions with resistor combinations, remembering that for parallel capacitors the capacitances add and for series capacitors the reciprocals of the capacitances add to give the reciprocal of the overall capacitance.

Calculation for capacitors in series

Once charging is complete, the charge on each capacitor is Q.

The p.d. across C_1 is given by $V_1 = \dfrac{Q}{C_1}$

The p.d. across C_2 is given by $V_2 = \dfrac{Q}{C_2}$

$V = V_1 + V_2 \therefore V = \dfrac{Q}{C_1} + \dfrac{Q}{C_2}$

Hence we can write $\dfrac{1}{C} = \dfrac{V}{Q} = \dfrac{1}{C_1} + \dfrac{1}{C_2}$

$\dfrac{1}{C} = \dfrac{1}{C_1} + \dfrac{1}{C_2}$

Worked example

(a) Calculate the effective capacitance of the combination shown. **(1 mark)**

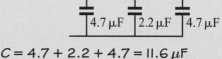

$C = 4.7 + 2.2 + 4.7 = 11.6\,\mu F$

(b) The charge on the system is $87\,\mu C$ when the combination is charged to a p.d. V. Calculate V. **(2 marks)**

$V = \dfrac{Q}{C} = \dfrac{87 \times 10^{-6}}{11.6 \times 10^{-6}} = 7.5\,V$

Now try this

A $560\,\mu F$ capacitor is connected in series with a $470\,\mu F$ capacitor. A 12 V battery is connected across the combination and the capacitors are allowed to charge fully. Calculate the energy stored in the combination. **(4 marks)**

Maths skills To simplify the arithmetic when calculating overall capacitance values, you can omit the standard prefix values ($\mu = 10^{-6}$, n $= 10^{-9}$ etc.) and use μF or nF as units, but remember that you will need to use the complete values when you calculate charge or p.d.

Capacitor circuits

Capacitors charge through a fixed resistor at a non-constant rate.

Charging through a fixed resistance

As the p.d. across the capacitor increases it opposes the e.m.f. and so the circuit current decreases.

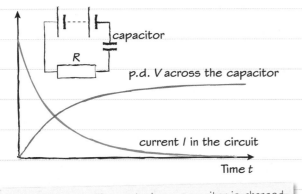

Change in p.d. and current when a capacitor is charged in a circuit in which resistance is constant

Discharging through a fixed resistance

The circuit shown can be used to charge and then discharge a capacitor. If the p.d. and current are recorded at regular intervals these graphs can be plotted.

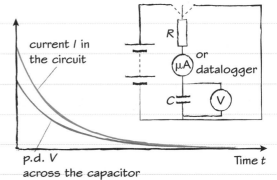

As the capacitor discharges through a resistor, the rate at which charge moves off the plates decreases over time. The current gradually decreases until the capacitor is fully discharged.

Time constant of a circuit

The time taken for the capacitor to become fully charged or discharged depends upon both the capacitance C and the circuit resistance R. A greater resistance will increase the time taken for charge to move onto or off the plate.

The product RC, known as the **time constant**, τ, of the circuit, is a key quantity for the charging and discharging processes. These are exponential processes and are covered on page 123.

In a time equal to one time constant a capacitor charges to 37% of full charge.

In one time constant the capacitor will discharge to 63% of its initial charge.

In a time equal to about 5 × RC a capacitor will be fully charged or discharged.

Charging a capacitor at a constant rate

Initially switch S is closed and the variable resistor is adjusted to set the ammeter to a convenient value.

When S is opened the capacitor starts to charge. As the p.d. across the capacitor increases, the p.d. across the variable resistor decreases.

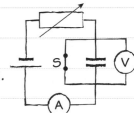

The variable resistor is adjusted to maintain the current at the chosen value. The charging current is monitored, and the p.d. across the capacitor is recorded at constant time intervals.

The procedure is repeated to obtain average values and a straight-line graph of p.d. across the capacitor against time can be plotted.

Worked example

Show that the time constant $\tau = RC$ has units equivalent to seconds. **(3 marks)**

R has units $\Omega = V A^{-1}$

C has units $F = C V^{-1} = A s V^{-1}$

RC has units $V A^{-1} A s V^{-1} = s$

Now try this

A 10 V power supply is used to charge a 4700 μF capacitor through a 2.2 kΩ resistor.

Calculate the time taken for the capacitor to charge fully. **(2 marks)**

Exponential processes

Capacitors charge and discharge exponentially through a fixed resistor.

Charging through a fixed resistance

Moving the switch from position 1 to position 2 begins discharging the capacitor.

If the switch in the circuit above is moved to 1, the capacitor charges through the fixed resistance. The charge on the capacitor will increase exponentially:

$$Q = Q_0(1 - e^{-\frac{t}{RC}})$$

where Q_0 is the final, maximum charge.

Again, the p.d. across the capacitor varies in the same way as the charge.

$$V = V_0(1 - e^{-\frac{t}{RC}})$$

The charging current decreases exponentially.

$$I = I_0 e^{-\frac{t}{RC}}$$

Exponential decay

A capacitor discharging through a fixed resistance is an example of **exponential decay**.

Exponential changes are constant-ratio changes. This means that the time taken for the charge on the capacitor to fall to a given fraction of the initial value is always the same for a given circuit.

The time t depends upon the capacitance, C, and resistance, R, in the circuit.

$$Q = Q_0 e^{-\frac{t}{RC}}$$

Q_0 is the initial, maximum charge on the capacitor and Q is the charge remaining on the capacitor after a time t. Note that $RC = \tau$, the time constant.

Similarly, during discharging, p.d. V and current I follow the expressions:

$$V = V_0 e^{-\frac{t}{RC}}$$
$$I = I_0 e^{-\frac{t}{RC}}$$

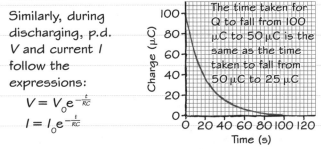

The time taken for Q to fall from 100 μC to 50 μC is the same as the time taken to fall from 50 μC to 25 μC

Modelling capacitor discharge

$$I = \frac{V}{R} = \frac{Q}{CR}, \text{ and } I = \frac{\Delta Q}{\Delta t}$$

$$\text{so } \Delta Q = \frac{Q}{CR} \times \Delta t$$

This formula can be used in a spreadsheet to calculate increments of charge ΔQ over time intervals Δt and thus to plot a graph modelling capacitor discharge.

Finding τ by a graphical method

$$Q = Q_0 e^{-\frac{t}{RC}}$$

Taking natural logarithms:

$$\ln Q = \ln Q_0 - \frac{t}{RC}, \text{ where } RC = \tau.$$

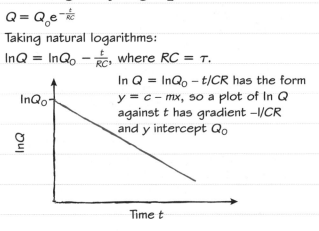

$\ln Q = \ln Q_0 - t/CR$ has the form $y = c - mx$, so a plot of $\ln Q$ against t has gradient $-1/CR$ and y intercept Q_0

Worked example

A 560 μF capacitor is charged to a p.d. of 230 V and then discharged through a circuit with resistance 68 Ω.

Calculate the time taken for the capacitor to discharge to a p.d. of 115 V. **(4 marks)**

$$V = V_0 e^{-\frac{t}{RC}} \therefore \frac{115}{230} = e^{-\frac{t}{68 \times 560 \times 10^{-6}}}$$

$$\therefore \ln(0.5) = -\frac{t}{0.0381} \therefore t = 0.0264 \text{s}$$

Maths skills Use logs to base e (natural logs) rather than logs to base 10. Your calculator has a $\log_e$ or ln button. The time must be in seconds for the calculation to be correct.

Now try this

Power supplies with an a.c. input have large-value capacitors to help keep the d.c. output voltage constant. In one power supply, the output voltage falls to zero every 10 ms. If the output must be maintained at a value at least 85% of its maximum value when a load of 2.2 kΩ is connected to the output terminals, show that the minimum capacitance needed in the power supply is 28 μF. **(3 marks)**

Exam skills

This exam-style question uses knowledge and skills you have already revised. Have a look at pages 120–123, for a reminder about capacitors.

Worked example

A defibrillator is an electrical device sometimes used to provide an electric shock to restart the heart when a patient's heart has stopped beating. The key component is a capacitor. The simplified circuit diagram below shows how a defibrillator can be charged from a power supply and then discharged through the patient.

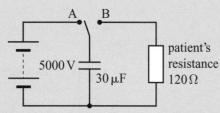

(a) Calculate the energy stored on the capacitor when it is fully charged. **(2 marks)**

$W = \frac{1}{2}CV^2 = \frac{1}{2} \times 30 \times 10^{-6} \times 5000^2$

$\qquad = 375\,J$

(b) Calculate the initial discharge current through the patient when the switch is moved from A to B. **(2 marks)**

$I = \frac{V}{R} = \frac{5000}{120} = 42\,A$

(c) Calculate the time constant for the discharge circuit and explain its significance. **(3 marks)**

Time constant $\tau = RC = 120 \times 30 \times 10^{-6}$
$\qquad\qquad\qquad = 0.0036\,s$
$\qquad\qquad\qquad = 3.6\,ms.$

The time constant is the time taken for the charge on the capacitor (and therefore the discharge current) to fall to 1/e of its initial value (0.37 of its initial value).

(d) Sketch a graph to show how the discharge current varies with time from the moment the switch is moved from A to B. Show the values of current after two time constants. **(3 marks)**

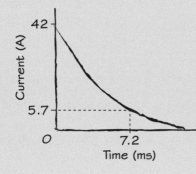

You might not be familiar with defibrillators, but an unfamiliar context should not put you off. The question is testing your knowledge and understanding of capacitor charging and discharging circuits.

The energy stored in a charged capacitor can be given by any of three expressions:
$$W = \frac{1}{2}CV^2 = \frac{1}{2}QV = \frac{1}{2}\frac{Q^2}{C}$$
It makes sense to use the equation which contains the information provided in the question: $W = \frac{1}{2}CV^2$.

Command word: Sketch

'Sketch' does not mean draw a vague diagram or artist's impression of a graph! You should still take care to:

☑ Draw the axes with a ruler.

☑ Label both axes with quantity and unit.

☑ Draw a smooth line or curve.

☑ Include values on the axes if you know them.

What you are not doing is plotting the graph. However, if there are known points that must lie on the graph, these can help you to draw it more accurately. In this case we know the time constant, so the current falls by a factor of 0.37 every 3.6 ms.

Note how the current at two time constants has been indicated. $2\tau = 7.2\,ms$ and the current is now $42 \times 0.37^2 = 5.7\,A$.

Electric fields

There is an electric field around every charged object. The interaction of two or more electric fields produces electrostatic forces.

Field lines for point and spherical charges

The field lines show us the direction in which a positive test charge experiences a force.
Field lines can therefore never cross.

The closer together the field lines are drawn, the stronger the electric field strength at that position.

The field around a point charge or a spherical charge distribution is radial.

The field around a point negative charge is directed radially inwards.

The field strength is zero inside any charged sphere. Outside the sphere the electric field varies as if there were a charge at its centre.

The field around a spherical positive charge distribution is directed radially outwards.

Interacting fields

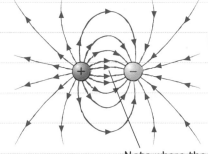

The field pattern when two opposite charges are brought close.

Note where the field is strongest for two opposite charges.

Forces on pairs of charged objects

For each pair of interacting charges, the two forces have the same lines of action (directions), are of the same type, are of equal magnitude, but act on different bodies. Hence the forces form an action–reaction pair (Newton's third law).

The force between like charges is repulsive.

The force between opposite charges is attractive.

Worked example

Two identical conducting spheres of mass $2.9\,m$ are hung from a fixed point by two threads with the spheres initially in contact. Charge is transferred onto one of the spheres, and the spheres move apart and come to equilibrium as shown.

Explain why the spheres move apart, and calculate the force on each sphere. **(2 marks)**

The conducting spheres share charge, so each sphere has the same charge. Like charges repel. The electrostatic forces (F) on each charge are horizontally outwards, and equal and opposite.

The forces are in equilibrium, so we can construct a force triangle.

$F = mg\tan 13$

Now try this

Draw the electric field variation between two equal magnitude positive charges. **(2 marks)**

Coulomb's law

Force between two charges

The force F between two point charges or spherical charged objects obeys an inverse square law, **Coulomb's law**.

$$F = \frac{kQq}{r^2}$$

where Q and q are the magnitudes of the two charges, r is the separation of the centres of the two charged objects and k is a constant dependent on the medium between the charges.

$$k = \frac{1}{4\pi\varepsilon_0}$$

where ε_0 is the permittivity of free space $\varepsilon_0 = 8.85 \times 10^{-12}\,\text{F m}^{-1}$, so $k = \frac{1}{4\pi\varepsilon_0} = 8.99 \times 10^9\,\text{m F}^{-1}$.

So $F = \dfrac{Qq}{4\pi\varepsilon_0 r^2}$

Electric field strength

We define the **electric field strength** E at a point as the force F acting per unit charge Q placed at that point:

$$E = \frac{F}{Q}$$

E has units N C^{-1}.

Electric field strength for a point charge

For a radial field produced by a point charge Q

$$E = \frac{Q}{4\pi\varepsilon_0 r^2}$$

E is a vector quantity and its direction is the same as that of the force.

Worked example

Two conducting spheres are placed with their centres 6.5 cm apart.

(a) Calculate the electrostatic force produced when the charge on one sphere is $+0.45\,\mu\text{C}$ and the charge on the other sphere is $+0.75\,\mu\text{C}$. **(2 marks)**

$$F = \frac{Qq}{4\pi\varepsilon_0 r^2} = \frac{7.5 \times 10^{-7} \times 4.5 \times 10^{-7}}{4\pi \times 8.85 \times 10^{-12} \times (6.5 \times 10^{-2})^2}$$

$\therefore F = 0.72\,\text{N}$

(b) Show that the field strength is zero at a point between the spheres about 3.7 cm from the sphere carrying the larger charge. **(4 marks)**

At this point r the field strength due to one sphere must be equal but oppositely directed to that produced by the other sphere.

$$E = \frac{Q}{4\pi\varepsilon_0 r^2}$$

$$\frac{7.5 \times 10^{-7}}{4\pi\varepsilon_0 r^2} = \frac{4.5 \times 10^{-7}}{4\pi\varepsilon_0 \times (6.5 - r)^2}$$

$$(6.5 - r)^2 = \frac{4.5 r^2}{7.5} = 0.6 r^2$$

This leads to the quadratic equation: $0.4 r^2 - 13 r + 42.25 = 0$

Substituting $r = 3.7$ gives $5.48 - 48.10 + 42.25 = -0.37$

which is near enough to 0 to suggest that r is about 3.7 cm. This can be confirmed by substituting $r = 3.6$ and $r = 3.8$ in turn.

Now try this

A van de Graaff generator has a spherical metallic dome of diameter 45 cm. When it has been running for some time the electric field strength at the dome's surface is $350\,\text{N C}^{-1}$. Calculate the charge on the dome. **(2 marks)**

Similarities between electric and gravitational fields

Electric and gravitational fields show a number of common features.

Gravitational field

Newton's law of gravitation for the force of attraction between two objects:

$$F = \frac{GMm}{r^2}$$

where

- F is the force of attraction between two objects
- G is the universal gravitational constant
- M and m are the masses of the objects
- r is the separation between the centres of the objects.

Both interactions obey an inverse square law.

Both interactions have an infinite range.

Gravitational field strength is the gravitational force per unit mass.

Gravitational forces act between masses.

Gravitational forces are always attractive.

The gravitational interaction is much weaker than the electrostatic interaction.

Electrostatic field

Coulomb's law for the force (of attraction or repulsion) between charges:

$$F = \frac{kQq}{r^2}$$

where

- F is the force of attraction or repulsion between two electrically charged particles
- k is the Coulomb force constant
- Q and q are the magnitudes of the point charges
- r is the separation between the particles.

Both interactions obey an inverse square law.

Both interactions have an infinite range.

Electrical field strength is the electrical force per unit charge.

Electrostatic forces act between charges.

Electrostatic forces can be attractive or repulsive.

The electrostatic interaction is much stronger than the gravitational interaction.

Worked example

Draw graphs to show the variation in gravitational field strength and gravitational potential around a sphere of uniform density. **(4 marks)**

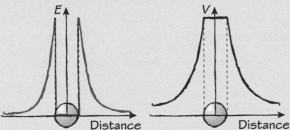

Electric field strength   Electric potential

The electric field strength and electric potential around a positively charged conducting sphere. Charge distributes itself until there is no potential difference within the conductor.

Now try this

A student is struggling with the concept of a field in physics. Making reference to electric and gravitational fields explain as clearly as you can how physicists use the field model. **(6 marks)**

Hence the electric potential is constant within the conducting sphere. Since there is no potential difference within the sphere, then the electric field strength must be zero within the sphere (as the magnitude of the electric field strength is equal to the potential gradient).

Uniform electric fields

As well as radial electric fields around point charges and charged wires, you need to know about uniform electric fields between parallel plates.

Field lines

The electric field lines of a **uniform field** between two charged plates are **parallel** and **equally spaced**.

The strength of the uniform field between the plates is given by:

$$E = \frac{V}{d}$$

where V is the **potential difference** between the plates and d is their separation.

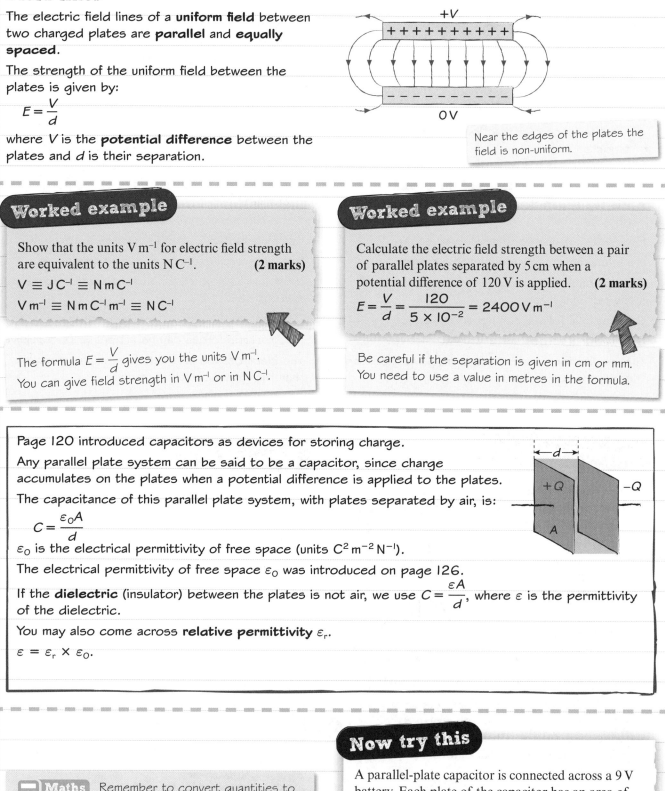

Near the edges of the plates the field is non-uniform.

Worked example

Show that the units $V\,m^{-1}$ for electric field strength are equivalent to the units $N\,C^{-1}$. **(2 marks)**

$V \equiv J\,C^{-1} \equiv N\,m\,C^{-1}$

$V\,m^{-1} \equiv N\,m\,C^{-1}\,m^{-1} \equiv N\,C^{-1}$

The formula $E = \frac{V}{d}$ gives you the units $V\,m^{-1}$.
You can give field strength in $V\,m^{-1}$ or in $N\,C^{-1}$.

Worked example

Calculate the electric field strength between a pair of parallel plates separated by 5 cm when a potential difference of 120 V is applied. **(2 marks)**

$$E = \frac{V}{d} = \frac{120}{5 \times 10^{-2}} = 2400\,V\,m^{-1}$$

Be careful if the separation is given in cm or mm. You need to use a value in metres in the formula.

Page 120 introduced capacitors as devices for storing charge.

Any parallel plate system can be said to be a capacitor, since charge accumulates on the plates when a potential difference is applied to the plates.

The capacitance of this parallel plate system, with plates separated by air, is:

$$C = \frac{\varepsilon_0 A}{d}$$

ε_0 is the electrical permittivity of free space (units $C^2\,m^{-2}\,N^{-1}$).

The electrical permittivity of free space ε_0 was introduced on page 126.

If the **dielectric** (insulator) between the plates is not air, we use $C = \frac{\varepsilon A}{d}$, where ε is the permittivity of the dielectric.

You may also come across **relative permittivity** ε_r.

$\varepsilon = \varepsilon_r \times \varepsilon_0$.

Now try this

Maths skills Remember to convert quantities to S.I. units. Be careful when converting area: $1\,cm^2 = 1 \times 10^{-4}\,m^2$.

A parallel-plate capacitor is connected across a 9 V battery. Each plate of the capacitor has an area of $16\,cm^2$, and the plates are separated by 5 μm of air.

Calculate the capacitance of this capacitor and the charge on each plate. **(4 marks)**

Charged particles in uniform electric fields

Charged particles experience forces in electric fields, so their motion may be changed.

Magnitude and direction

A charge q in a region of electric field strength E experiences a force of magnitude Eq. The sign of the charge determines the direction of the force.

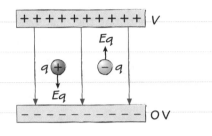

Deflection

When a charged particle enters a region of electric field it will experience a force. If the particle enters perpendicularly to the electric field, then the acceleration will act at right angles to the original direction of travel. The particle experiences motion at a constant speed in one direction, and uniform acceleration in a perpendicular direction. This causes **projectile motion**, and the object follows a **parabolic path**, resembling the motion of a projectile that has gravitational acceleration vertically and a constant velocity horizontally.

Remind yourself about projectile motion on page 14.

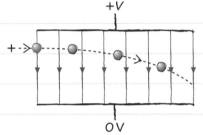

Worked example

An electron travelling horizontally at a speed of $1.5 \times 10^7 \, \text{m s}^{-1}$ enters a region between two horizontal parallel plates, 2 cm apart and 6 cm long. The potential difference between the plates is 80 V.

Calculate the deflection from the horizontal of the electron when it leaves the region. **(5 marks)**

Time in region, $t = \dfrac{0.06}{1.5 \times 10^7}$

$= 4.0 \times 10^{-9} \, \text{s}$

Field strength, $E = \dfrac{V}{d}$

$= \dfrac{80}{0.02}$

$= 4000 \, \text{V m}^{-1}$

Acceleration, $a = \dfrac{F}{m}$

$= \dfrac{qE}{m}$

$= \dfrac{1.6 \times 10^{-19} \times 4000}{9.11 \times 10^{-31}}$

$= 7.025 \times 10^{14} \, \text{m s}^{-2}$

$s = ?, \, u = 0, \, t = 4.0 \times 10^{-9} \, \text{s},$

$a = 7.025 \times 10^{14} \, \text{m s}^{-2}$

Deflection $s = ut + \frac{1}{2}at^2$

$= 0 + \frac{1}{2} \times 7.025 \times 10^{14} \times (4.0 \times 10^{-9})^2$

$= 5.6 \times 10^{-3} \, \text{m or } 5.6 \, \text{mm (2 s.f.)}$

① Use the initial speed of the particle and the length of the region to find the time the particle spends in the field.

② Find the particle's acceleration due to the electric field. $F = qE$ and $F = ma$, so $a = \dfrac{qE}{m}$.

③ Find the deflection of the particle using the equation $s = ut + \frac{1}{2}at^2$.

If you have to use one of the constant acceleration formulae (SUVAT) it's a good idea to write down the variables you know and the variable you want to find. This helps you choose the correct formula.

Remind yourself about SUVAT equations on page 11.

Now try this

An electron travelling horizontally at a speed of $1.2 \times 10^7 \, \text{m s}^{-1}$ enters a region between two horizontal parallel plates, 5.0 cm long. The electric field strength between the plates is $3600 \, \text{V m}^{-1}$.

Calculate the deflection from the horizontal of the electron when it leaves the region. **(4 marks)**

Electric potential and electric potential energy

Charged particles experience forces in electric fields, so work may be done as their position in the field changes.

Doing work

The charge $+Q$ in the region of uniform electric field below experiences a force in the direction of the field. As the charge moves from position 1 to position 2 **work** is done on the charge.

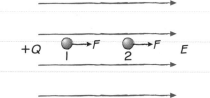

To move the charge from position 2 to position 1 the charge must do work.

When work is done there is an energy transfer.

Electric potential difference

The kinetic energy of a positively charged particle increases when it moves in the direction of an electric field.

Energy must be **conserved**, and so there must be a decrease in another store of energy as the kinetic energy increases. We call this energy store the **electric potential energy**.

The work done, W, per coulomb of charge moved between two points in the field is called the **electric potential difference**, V, between the two points.

$$V = \frac{W}{Q}$$

Electric potential in a radial field

Work must be done by a positive charge when it is brought from infinity to a point in an electric field.

The **potential** is defined to be zero at infinity. Hence the potential at the final position of the charge is equal to the work done per unit charge.

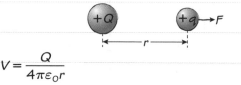

$$V = \frac{Q}{4\pi\varepsilon_0 r}$$

Potential is a **scalar** quantity – it has a magnitude but no direction.

A positive charge produces a positive potential and a negative charge produces a negative potential.

To find the resultant potential at a point due to a number of charges we simply add the magnitudes of the individual potentials (taking the sign into account).

Electric potential energy

The work done when a charge q is moved between two points in the field of a point charge is $W = Vq$. Since V is defined to be zero at infinity, the electric potential energy E of a charge moved to the point charge Q from infinity is $E = W = Vq$.

$$V = \frac{Q}{4\pi\varepsilon_0 r}$$

Be careful not to confuse E for electric potential energy with E for electric field strength.

So the electric potential energy $E = \dfrac{Qq}{4\pi\varepsilon_0 r}$ at a distance r from the point charge Q.

Worked example

A hydrogen nucleus consists of a single proton. Bohr estimated the diameter of the hydrogen atom to be 1.05×10^{-10} m.

(a) Calculate the electric potential produced by the proton at a distance 5.3×10^{-11} m from its centre. **(2 marks)**

$$V = \frac{Q}{4\pi\varepsilon_0 r}$$

$$= \frac{1.6 \times 10^{-19}}{4\pi \times 8.85 \times 10^{-12} \times 5.3 \times 10^{-11}}$$

$$= 27\,V$$

(b) Hence determine the energy required to remove an electron to infinity from this position. **(2 marks)**

$$W = QV = 1.6 \times 10^{-19} \times 27 = \underline{4.3 \times 10^{-18}\,J}$$

Now try this

(a) Two small, positively charged oil droplets are held at rest at a separation of 2 cm. Each droplet carries a charge of $1\,\mu C$. The droplets are released and the charges start to accelerate away from each other. Calculate the net kinetic energy of the system when the charges are very far from one another. **(2 marks)**

(b) Explain why the kinetic energy would only be shared equally between the two droplets if they had the same mass. **(2 marks)**

Capacitance of an isolated sphere

An isolated charged sphere stores energy like a capacitor.

Charge on a sphere

The electric potential outside a charged conducting sphere, a distance r from its centre, is:

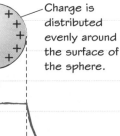

Charge is distributed evenly around the surface of the sphere.

$$V = \frac{Q}{4\pi\varepsilon_0 r}$$

The electric potential around a conducting sphere. Within the sphere the electric potential is constant and so the field strength is zero.

Capacitance of a sphere

If charge is placed on an isolated sphere, then the charge will be stored. The sphere acts as a capacitor.

On page 120, capacitance, C, was defined as the charge stored, Q, per unit potential difference, V.

$$C = \frac{Q}{V}$$

Substituting in the electric potential V outside a conducting sphere: $C = \frac{4\pi\varepsilon_0 rV}{V}$, so the capacitance of an isolated sphere is

$C = 4\pi\varepsilon_0 R$, where R is the radius of the sphere.

Work is done as charge is placed on the sphere. When the sphere is discharged there will be an energy transfer. Hence we can say that the sphere stores energy because it stores charge.

Worked example

(a) Show that a charge of about 9 nC must be placed on a conducting sphere of radius 5.0 cm for the electric potential at the surface of the sphere to be 1600 V. **(2 marks)**

$Q = 4\pi\varepsilon_0 RV$

$Q = 4\pi \times 8.85 \times 10^{-12} \times 5.0 \times 10^{-2} \times 1600$

$\quad = 8.9 \times 10^{-9}\,C$

(b) When this charge is placed on the sphere, what is the electric potential at a distance of (i) 10.0 cm, (ii) 2.5 cm from the centre of the sphere? **(3 marks)**

(i) $V \propto \frac{1}{r}$, r doubles from 5.0 to 10.0 cm,

so V halves: $V = 800\,V$

(ii) $V = 1600\,V$ ($r = 2.5$ cm is inside the sphere, and potential is constant throughout the sphere.)

Work done on a moving charge

For a charge being moved in a repulsive radial electric field around a charged sphere, the force varies with distance.

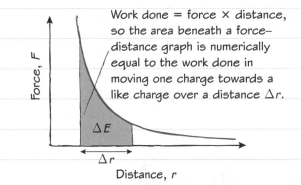

Work done = force × distance, so the area beneath a force–distance graph is numerically equal to the work done in moving one charge towards a like charge over a distance Δr.

The same principle applies for the work done by a charge moving in an attractive radial field.

Now try this

The positive lead from a high-voltage power supply is momentarily connected to a conducting sphere of radius 7.5 cm, raising the sphere to a potential of 1800 V.

(a) Calculate the capacitance of the sphere, and the charge on its surface ($\varepsilon_0 = 8.85 \times 10^{-12}\,C^2\,N^{-1}\,m^{-2}$). **(4 marks)**

(b) The sphere is connected to an identical sphere, which is initially uncharged.
Calculate the potential at the surface of each sphere after the two spheres are connected. **(3 marks)**

Hint: When the connection is made, charge must be conserved, and the two spheres must both have the same potential at their surfaces.

Representing magnetic fields

A magnetic field is a region in which a magnetic pole or an electric current experiences a force.

Magnets

Some materials, such as iron, nickel, and cobalt, can be magnetised when placed in a region of magnetic field.

Some materials are magnetically hard and retain their magnetism once magnetised.

The magnetic field around a magnet can be revealed using iron filings, since the iron filings become magnetised and experience magnetic forces.

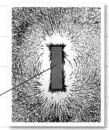

bar magnet

Attraction and repulsion

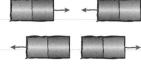

Unlike poles attract and like poles repel.

The field increases between like poles as the poles are brought closer together.

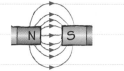

There is a neutral point between two like poles.

Lines of magnetic field

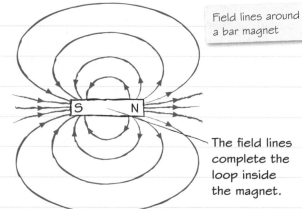

Field lines around a bar magnet

The field lines complete the loop inside the magnet.

Like electric field lines, magnetic field lines never cross. They represent the force that would act on a north (N) pole, so their direction is from N to S.

The closer together the lines are in a given region the greater the strength of the magnetic field.

Magnetic fields around conductors

Moving charges create their own magnetic field, so current-carrying conductors experience a force in a magnetic field.

The right-hand rule: point your thumb in the direction of the conventional current and your fingers will curl in the direction of the field lines.

A flat coil made of one loop of current-carrying wire produces a field like this.

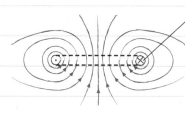

A circle with a dot represents current emerging from the plane of the diagram, while a circle with a cross means that the current is moving into the plane.

A solenoid produces a field similar to that of a bar magnet.

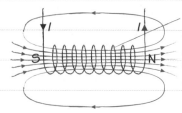

Use the right-hand rule on one of the coils going into or emerging from the diagram to see which way the field lines point.

Now try this

Explain why the field lines in a diagram representing a magnetic field should never cross.　　　**(2 marks)**

Force on a current-carrying conductor

The force on a current-carrying conductor acts at right angles to the field and to the current direction.

Magnetic flux density

If the current, I, is along the direction of the magnetic field, B, then the magnetic force is zero.

If the current is perpendicular to the direction of the magnetic field then the magnetic force is a maximum.

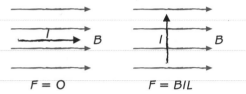

$$F = 0 \qquad F = BIL$$

$F = BIL$, where L is the length of conductor.

This equation defines the **magnetic flux density**, B, as the force per unit current element. B has units $N\,A^{-1}\,m^{-1} = T$, **tesla**.

Force on a charge

Similarly the force on a charge Q moving with a velocity v perpendicular to the direction of a magnetic field is a maximum.

$$F = BQv$$

If a beam of negatively charged particles passes through a magnetic field, the force direction is as if positively charged particles were moving in the opposite direction.

For charge movement at an angle θ to the field, the component at right angles to the field is used, modifying the force by a factor $\sin\theta$.

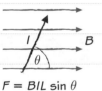

$$F = BIL\sin\theta$$

This gives the equations:

$$F = BIL\sin\theta \quad \text{and} \quad F = BQv\sin\theta$$

Fleming's left-hand rule

The magnetic force is perpendicular to both the current and the magnetic field direction.

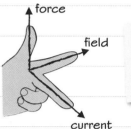

The direction of the magnetic force can be identified by applying **Fleming's left-hand rule**. In this rule, I is always the conventional current.

Determining uniform magnetic flux density

A current-carrying wire is fixed between the poles of a magnet placed on a digital balance. When the wire carries a current, a force will be exerted on the wire in accordance with Fleming's left-hand rule. The wire exerts an equal and opposite force on the magnet, changing the reading on the balance. The size of this force is calculated from $F = mg$ for a range of values of the current I, so the magnetic flux density can be calculated from $F = BIL$.

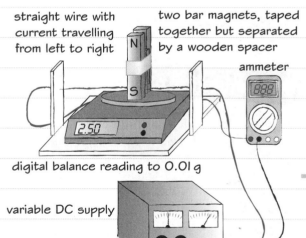

straight wire with current travelling from left to right

two bar magnets, taped together but separated by a wooden spacer

ammeter

digital balance reading to 0.01 g

variable DC supply

Worked example

A conductor carries a current of 155 mA at an angle of 40° to a field of strength 0.15 T.

(a) Calculate the force on a 25 cm length of this conductor. **(2 marks)**

$$F = BIL\sin\theta = 0.15 \times 0.155 \times 0.25 \times \sin40$$
$$= 3.74 \times 10^{-3}\,N$$

(b) At what angle would the force on this length be half as big as that in (a)? **(2 marks)**

$$\sin\theta = \frac{0.5 \times 3.74 \times 10^{-3}}{0.15 \times 155 \times 10^{-3} \times 0.25} = 0.322$$
$$\therefore \theta = \sin^{-1}(0.322) = 18.8°$$

Now try this

Electrons are emitted from an electron gun in a cathode ray tube at a speed of $1.2 \times 10^7\,m\,s^{-1}$. They enter a magnetic field of strength 140 mT perpendicularly, and are deflected.

Calculate the acceleration of the electrons when they enter the field. **(3 marks)**

Charged particles in a region of electric and magnetic field

Charged particles may be deflected into circular paths when passing through a region of uniform magnetic field.

Deflection in a magnetic field

When charged particles enter a region of uniform magnetic field at right angles to the field, they experience a magnetic force F, which is at right angles to the direction of the charge movement.

The magnetic force is given by $F = BQv$

F causes a centripetal acceleration, $a = \dfrac{v^2}{r}$ and the particles are deflected into a circular path of radius r.

$$BQv = \frac{mv^2}{r}$$

The radius of the circular path r is given by $r = \dfrac{mv}{BQ}$

magnetic field
B acts into
paper ⊙ ⊙ ⊙ ⊙ ⊙ ⊙

Negatively charged particles experience a force in the direction of the force on positively charge particles moving in the opposite direction.

Deflection in an electric field

When charged particles enter a region of uniform electric field they are deflected into a parabolic path, as explained on page 129.

There is more about projectile motion on page 14, and page 129 looks at the effect for charged particles.

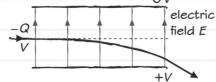

Negatively charged particles experience a force in the opposite direction to the field, and are deflected towards the positive plate.

Magnetic and electric fields

A B field applied directly inwards can be adjusted to give a force equal but opposite to the electric force, leading to a straight path for the particle. If particles of a range of velocities enter the combined fields, the fields can be adjusted so that only particles with a particular velocity will emerge in a straight line. This **velocity selector** is used in mass spectrometry.

Worked example

A uniform magnetic field of flux density 0.575 mT is used to return the electrons in the electron beam tube of the previous question to a straight-line path. Calculate the p.d. applied between the plates. **(2 marks)**

$EQ = BQv$

$\therefore E = 0.575 \times 10^{-3} \times 1.1 \times 10^7 = 6330\,V\,m^{-1}$

$E = \dfrac{V}{d}$

$\therefore V = 6330 \times 7.5 \times 10^{-2} = 470\,V$

Worked example

Electrons are accelerated in the 'gun' of an electron beam tube through a p.d. of 350 V. They are deflected into a parabolic path as they pass through a pair of parallel plates that are 7.5 cm apart.

Show that the speed of the electrons as they leave the 'gun' is about $1 \times 10^7\,ms^{-1}$. **(2 marks)**

$\frac{1}{2}mv^2 = QV$

$\therefore v = \sqrt{\dfrac{2 \times 1.6 \times 10^{-19} \times 350}{9.1 \times 10^{-31}}} = 1.1 \times 10^7\,ms^{-1}$

Now try this

(a) It is observed that when an electron beam enters a uniform magnetic field at right angles to the field the beam is deflected into a circular path. Explain this observation. **(3 marks)**

(b) It is observed that when the electron beam enters the same field at an angle close, but not equal, to 90° to the field the beam is deflected into a helical path. Explain this observation. **(3 marks)**

Magnetic flux and magnetic flux linkage

Magnetic flux and flux linkage are useful ways to quantify the effect of a magnetic field.

Magnetic flux

The closeness of the magnetic field lines around a magnet or current-carrying conductor in a diagram indicates the field strength. The closer together the field lines are in a given region, the larger the **flux density**, B.

Magnetic flux, ϕ, through a plane surface is defined as:

$\phi = BA$ (units $T\,m^2 = Wb$, **weber**).

where B is the flux density perpendicular to the surface and A the area of the surface.

$\phi = B.A$

> Magnetic flux is measured perpendicular to the plane.

Magnetic flux calculation

If the field is at an angle to the surface, then the component of B perpendicular to the plane of the surface is taken.

$\phi = BA\cos\theta$

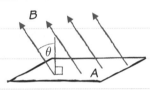

> ϕ is a maximum when $\theta = 0$ and is zero when $\theta = 90°$.

Magnetic flux linkage

We often need to calculate the flux that is linked with a conductor.

Compare a one-turn coil and a coil of N turns. Each extra turn increases the flux linked between the magnet and the coil.

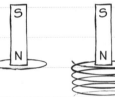

1 turn: ϕ N turns: $N\phi$

Magnetic flux linkage for a coil is the product of the magnetic flux through the coil and the number of turns on the coil ($N\phi$, units weber-turns).

Worked example

A student is confused about the differences between magnetic flux, magnetic flux density and magnetic flux linkage. Explain as clearly as you can the differences between these quantities. In your explanation include a discussion of how these quantities are linked. **(4 marks)**

The magnetic flux ϕ is the total amount of magnetic field lines in a given region. The magnetic flux density B is the amount of flux passing through a unit area A at right angles to the magnetic field. This is a measure of the strength of the field.

The flux density is linked to the magnetic flux by the equation $B = \phi/A$.

The magnetic flux linkage $N\phi$ for a coil is the product of the magnetic flux through the coil and the number of turns N on the coil.

Worked example

A solenoid is made by winding insulated copper wire around a wooden dowel of circular cross-section. The diameter of the dowel is 7.5 cm.

A current is passed through the solenoid, and a uniform field of flux density 4.8 mT is produced inside the solenoid. Calculate the flux linked with one turn of the solenoid. **(2 marks)**

$\phi = B.A = 4.8 \times 10^{-3} \times \pi\left(\dfrac{7.5 \times 10^{-2}}{2}\right)^2$

$\therefore \phi = 2.1 \times 10^{-5}\,Wb$

Now try this

The flux linked with a flat coil whose plane is perpendicular to the magnetic field is 1.3×10^{-3} Wb. Calculate the minimum angle that the plane of the coil would have to rotate through in order for the flux linkage to decrease to 7.5×10^{-4} Wb. **(2 marks)**

Faraday's law of electromagnetic induction and Lenz's law

An e.m.f. is induced whenever there is a change in flux linkage associated with a conductor.

Induced e.m.f.

When a conductor experiences a change of flux linkage, an e.m.f. is induced in the conductor.

The change in flux linkage may be due to relative motion between the conductor and the field, or due to magnetic flux density changes.

Faraday discovered the effect, which is summed up in the laws of electromagnetic induction.

Inducing a current

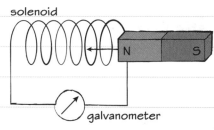

solenoid

galvanometer

When the magnet is moved into or out of the solenoid, the galvanometer deflection shows that an induced current flows.

The faster the magnet is moved, the larger the deflection, indicating that the rate of change of flux linkage determines the induced e.m.f.

If the direction the magnet is moved is changed, the current direction also changes.

A N pole moved towards one end of the solenoid induces a current producing a N pole at that end; moving a N pole away induces a S pole.

The laws of electromagnetic induction

Faraday's law: The magnitude of the induced e.m.f. is proportional to the rate of change of flux linkage associated with the conductor.

Lenz's law: The e.m.f. is induced in a direction so as to oppose the change producing the e.m.f.

The laws can be summarised as $\varepsilon = -\dfrac{\Delta(N\phi)}{\Delta t}$

Eddy currents

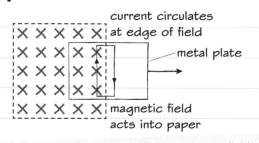

current circulates at edge of field

metal plate

magnetic field acts into paper

Moving a metal plate through the edge of a field

As the metal plate is moved through the magnetic field a current flows at the edge of the field. This is because an e.m.f. is induced within, but not outside of, the field.

Worked example

A copper wire of length L is moved at velocity v horizontally through a uniform vertical magnetic field of flux density B. Show that the magnitude of the induced e.m.f. is $\varepsilon = BLv$.

Calculate the minimum speed required for a wire of length 7.5 m to have an induced e.m.f. of 375 mV when it is moving through a field of flux density 52 μT. **(4 marks)**

Area swept out per second by wire is vL.

$$\frac{\Delta(N\phi)}{\Delta t} = B\frac{\Delta A}{\Delta t} = BvL$$

$$\varepsilon = BvL \qquad v = \frac{375 \times 10^{-3}}{52 \times 10^{-6} \times 7.5} = 960 \text{ ms}^{-1}$$

Worked example

A magnet is dropped through a coil connected to a datalogger. Explain the shape of the graph. **(2 marks)**

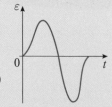

As the magnet approaches the coil, the flux linkage increases and an e.m.f. is induced. There is a change in direction of the e.m.f. because the poles of the magnet 'swap' location relative to the coil as it passes through the centre.

Once it has passed through the coil, flux linkage decreases and the e.m.f. changes direction.

The magnet has been accelerating so the magnitude of the negative peak is larger and the time taken for the flux linkage to become zero is reduced.

Now try this

When a conductor experiences a change in flux linkage, an e.m.f. is induced in a direction so as to oppose the change producing the e.m.f. Explain why the direction of the e.m.f. must be so as to oppose the change producing it. **(3 marks)**

The search coil

When the flux linked with a coil of wire changes, an e.m.f. is induced.

Search coil

A search coil is a small coil with a large number of turns. When the coil is placed in a region of varying magnetic field, an e.m.f. is induced in the coil. The e.m.f. can be measured by connecting the search coil to a cathode ray oscilloscope (CRO) or datalogger.

For a given search coil, the maximum induced e.m.f. ε_0 is proportional to the maximum magnetic flux density B_0 as long as the frequency of the field variation is constant.

Use of a search coil

If a sinusoidal a.c. of frequency f is applied,
$N\phi \propto B_0 \sin(2\pi ft)$

$$\varepsilon = -\frac{\Delta(N\phi)}{\Delta t} \propto fNB_0 \sin(2\pi ft) \therefore \varepsilon_0 \propto fNB_0$$

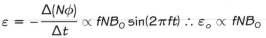
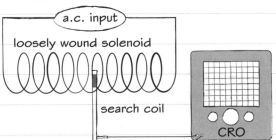

Generating an e.m.f.

If a coil is rotated in a magnetic field as shown, an e.m.f. is generated as a result of the changes in flux linkage associated with the coil.

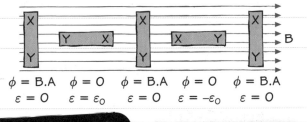

A simple a.c. generator.

The rate of change of flux linkage is largest when the coil is along the field direction, and least when the coil is perpendicular to the field.

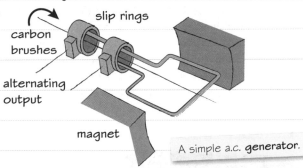

$\phi = B.A$ $\phi = 0$ $\phi = B.A$ $\phi = 0$ $\phi = B.A$
$\varepsilon = 0$ $\varepsilon = \varepsilon_0$ $\varepsilon = 0$ $\varepsilon = -\varepsilon_0$ $\varepsilon = 0$

Value and phase of e.m.f.

There is a phase difference of $\frac{\pi}{2}$ rad between the flux linked with the coil and the induced e.m.f. (The induced e.m.f. lags behind the flux linkage by a quarter of a cycle.)

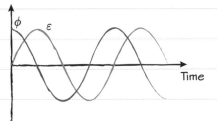

An alternating p.d. may also be generated by rotating a magnetic field within a stationary coil.

The value of the maximum induced e.m.f. depends on:
- the number of turns in the coil
- the strength of the field
- the speed of rotation.

Worked example

A 2200-turn search coil of area 1.25 cm² is placed between the pole pieces of a magnet. The flux density between the poles is 0.35 T. The flux is perpendicular to the plane of the coil. When the coil is removed from the field the average e.m.f. induced is 180 mV. Calculate the time taken for the flux linkage to decrease to zero. **(3 marks)**

$N\phi = NBA = 2200 \times 0.35 \times 1.25 \times 10^{-4}$

$= 0.096 \, \text{Wb}$

$\varepsilon = -\frac{\Delta(N\phi)}{\Delta t} \therefore t = \frac{0.096}{180 \times 10^{-3}} = 0.53 \, \text{s}$

Now try this

The graph shows how the flux linkage varies with time for a coil rotating in a uniform magnetic field.

Explain the variation in the induced e.m.f. with time over the time interval indicated. **(4 marks)**

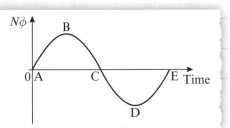

The ideal transformer

A changing current in a conductor can induce an e.m.f. in another conductor placed nearby.

The transformer effect

A changing current produces a region of changing magnetic field. This induces an e.m.f. in another conductor placed in this region.

A pair of magnetically linked coils is called a **transformer**. The coil carrying the current is called the primary coil. The other coil is called the secondary coil.

In a practical transformer, magnetic flux produced by the primary coil is linked to the secondary coil by means of a soft (easily magnetised) iron core.

Turn numbers and potential difference

For the primary coil, where n_p is the number of turns, $\varepsilon_p = -\dfrac{\Delta(n_p\phi_p)}{\Delta t}$

$\varepsilon_p = -V_p \therefore V_p = \dfrac{\Delta(n_p\phi_p)}{\Delta t} = n_p\dfrac{\Delta\phi_p}{\Delta t}$

Similarly, for the secondary coil,

$\varepsilon_s = -\dfrac{\Delta(n_s\phi_s)}{\Delta t} = -n_s\dfrac{\Delta\phi_s}{\Delta t}$

When there is maximum flux linkage between the two coils,

$\dfrac{\Delta\phi_p}{\Delta t} = \dfrac{\Delta\phi_s}{\Delta t} \therefore \dfrac{\varepsilon_s}{V_p} = -\dfrac{n_s}{n_p}$

Conservation of energy

For an ideal transformer, one with no losses, the power in the secondary is the same as that in the primary. This is a consequence of energy conservation. $P_s = P_p$

$V_sI_s = V_pI_p$

$\dfrac{I_s}{I_p} = \dfrac{V_p}{V_s} \qquad \therefore \dfrac{I_s}{I_p} = \dfrac{n_p}{n_s}$

If the number of turns in the secondary coil n_s is greater than the number of turns in the primary n_p, we have a step-up transformer. The p.d. is stepped up, and the current is stepped down in the same ratio.

In reality other energy transfers take place too, such as ohmic heating in the transformer windings and eddy current losses in the core.

The windings are made of thick wire to minimise their resistance, and the core is laminated.

A laminated core consists of thin layers of iron assembled together, with electrical insulation between. This construction restricts the flow of eddy currents and so reduces ohmic heating losses.

A simple transformer

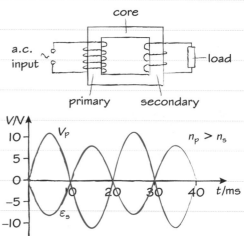

Investigating transformers

Construct a simple transformer by winding two coils of insulated copper wire around a soft iron core. Wind 10 turns for the primary and 25 turns for the secondary coil. Connect the wires from the primary coil to a low-voltage a.c. power supply set at 2 V. Keeping the number of turns on the primary coil constant, measure the p.d. across the primary and secondary coils with a multimeter set to a.c. as the number of coils on the secondary is changed. Use the same technique to investigate the effect of the number of turns in the primary coil.

When you change the numbers of turns of wire on each coil, the total length should remain constant so that its resistance does not change.

Worked example

There are 300 turns in the primary coil and 450 turns in the secondary coil of a transformer. Calculate the e.m.f. induced in the secondary coil when a 12 V a.c. is applied to the primary. **(2 marks)**

$\dfrac{V_s}{V_p} = \dfrac{n_s}{n_p} \qquad \therefore V_s = \dfrac{450}{300} \times 12 = 18\,V$

Now try this

A physics student suggests that in a step-up transformer the law of conservation of energy is being broken, since the e.m.f. induced in the secondary coil is larger than the e.m.f. applied to the primary coil. Explain why the suggestion is incorrect. **(3 marks)**

Exam skills

This exam-style question uses knowledge and skills you have already revised. Have a look at pages 135 and 136 for a reminder about magnetic flux linkage and Faraday's and Lenz's laws.

Worked example

The diagram shows a simple experiment used to demonstrate electromagnetic induction.

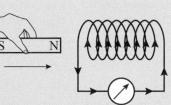

As the bar magnet is moved towards one end of the coil the ammeter indicates a small current flowing around the circuit.

(a) Explain how the current is generated. **(4 marks)**

The bar magnet creates a **magnetic field** in the space around it. As the bar magnet moves towards the coil, the **magnetic flux** through the coil changes.

Faraday's law states that there will be an **induced e.m.f.** in the coil equal to the **rate of change of magnetic flux linkage**.

The circuit is complete, so the **e.m.f.** results in a flow of **current**.

(b) State and explain the effect of reversing the direction of motion of the bar magnet and moving it away from the coil at a higher speed. **(4 marks)**

The direction of current in the circuit would reverse and the magnitude of the current would increase.

The current reverses because the magnet has changed direction. This changes the sign of the rate of change of flux linkage and so reverses the sign of the induced e.m.f.

The current increases in magnitude because the rate of change of flux linkage is greater.

(c) The current induced in the coil dissipates some energy in the circuit. Use Lenz's law to explain how energy is conserved in this experiment. **(5 marks)**

- The induced current makes the coil into an electromagnet that exerts forces on the bar magnet.
- Lenz's law states that the induced e.m.f. is in a direction such as to oppose the change that caused it.
- This means that the current is in such a direction that it exerts a repulsive force on the bar magnet when the magnet approaches and an attractive force on it when it is moved away.
- The person moving the magnet must do work against these forces.
- The energy transferred by this work is equal to the energy dissipated as heat in the circuit.

(d) When the bar magnet is moved towards the coil, the flux through the circuit changes by 8.0×10^{-4} Wb in 0.50 s. The coil has 20 turns and the total resistance of the circuit is $5.0\,\Omega$. Calculate the average induced current during this time. **(4 marks)**

Change of flux linkage = $8.0 \times 10^{-4} \times 20 = 0.016$ Wb

Rate of change of flux linkage = $\dfrac{0.016}{0.50} = 0.032$ Wb s^{-1}
$\qquad\qquad = $ induced e.m.f.

Average current = $\dfrac{\text{induced e.m.f.}}{\text{resistance}} = \dfrac{0.032}{5.0} = 6.4 \times 10^{-3}$ A
$\qquad\qquad\qquad = 6.4$ mA.

The introduction mentions electromagnetic induction – this should alert you to use Faraday's and Lenz's laws and to think about magnetic field strength and flux.

When explaining try to include relevant technical terms – this helps to show that you understand the important underlying physics.

Read the question carefully and make sure you answer all parts of it. In this question there are two things to state and two explanations to give.

An alternative approach to answering (b) would be to start by quoting Faraday's and Lenz's laws in the form:

$$E = -\dfrac{\Delta(N\phi)}{\Delta t}$$

and then indicating how the changes affect this.

This part asks you to use Lenz's law, so it is important to show how you have used it. The simplest way to do this is to state the law before you apply it.

Using bullet points for a question worth several marks helps to ensure that you are making separate points. It also sets up a logical structure to your answer.

Notice how this has been structured – whilst it would be acceptable to quote Faraday's law and substitute all the values straight into the equation, it is often helpful to break up the calculation into simple steps.

The nuclear atom

At the beginning of the twentieth century a series of experiments established the nuclear model of the atom.

Alpha particle scattering

The nuclear model of the atom was developed following the results of an experiment in which alpha particles were directed at thin gold foil.

The vast majority of the alpha particles experienced no or very little deflection as they passed through the foil, but a tiny fraction deflected through very large angles. Some alpha particles even rebounded back from the foil.

The nuclear model

α particle

gold nucleus

The **nuclear model** of the atom explains the alpha particles' deflection: an atom is made mostly of empty space but has a tiny, dense nucleus containing all of the positive charge in the atom and most of its mass. Nuclei have a radius of a few femtometres (10^{-15} m) whereas atomic radii are measured in tens or hundreds of picometres (1 pm = 10^{-12} m). If the nucleus of an atom were enlarged to the size of a tennis ball, its electrons would be a kilometre away.

The nucleus consists of protons and neutrons (collectively called nucleons). The **proton number** (or **atomic number**), Z, identifies the element, and it is also equal to the number of electrons surrounding the nucleus in the neutral atom. The **mass number** or **nucleon number**, A, is the total number of nucleons in the nucleus.

Notation

We use symbols to represent the type of a nucleus (nuclide) as:

$$^{A}_{Z}X$$

X is its chemical symbol.

A carbon nucleus normally consists of 6 protons and 6 neutrons, so we write

$$^{12}_{6}C$$

for the major isotope of carbon.

Isotopes

Nuclides with the same number of protons but different numbers of neutrons are called **isotopes**. Isotopes have different mass numbers, A. For example, carbon-14 is a radioactive isotope of carbon.

$$^{14}_{6}C$$

Although it has 6 protons in its nucleus, there are 8 neutrons. The extra neutrons lead to this isotope being unstable.

Worked example

Write the symbols for the following nuclides: aluminium, with 13 protons and 14 neutrons; sulfur, with 16 neutrons and 16 electrons. **(2 marks)**

Al: $Z = 13$, $A = 13 + 14 = 27$ so $^{27}_{13}Al$

S: number of electrons = number of protons in neutral atom so $Z = 16$;
$A = 16 + 16 = 32$ so $^{32}_{16}S$

In the neutral atom there will always be the same number of electrons as there are protons in the nucleus.

Light nuclei have about the same number of neutrons as protons, but in heavy nuclei the ratio of neutrons to protons is about 3 : 2.

Remember: You do not have to memorise the chemical symbols for elements, or their proton numbers.

Now try this

1 The figure shows the path of an alpha particle passing near a nucleus.

α particle　　　gold nucleus

(a) Add to the diagram to show the force on the alpha particle in the position where the force is a maximum.

(b) The nucleus is replaced with one that has a smaller proton number. Draw on the diagram the path of an alpha particle that starts with the same velocity and position as that of the alpha particle drawn. **(2 marks)**

2 Complete the table for the following nuclides:

Number of protons	Number of neutrons	Number of electrons	Symbol
2	2	2	$^{4}_{2}He$
8	8		O
	138	88	Ra
			$^{238}_{92}U$

(3 marks)

Nuclear forces

At the tiny distances in the nucleus, protons and neutrons experience the strong nuclear interaction.

The strong nuclear force

As protons are brought together, the electrostatic (coulomb) repulsive force acting on them grows larger and larger.

Gravitation is a weak interaction and cannot hold the protons together. Therefore there must be another very strong force acting between protons at nuclear separations. This force acts between all nucleons and is called the **strong nuclear force**.

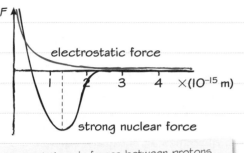

The variations in forces between protons over distance: positive values are repulsive.

Nuclear density

The strong nuclear force is very short range. At separations less than about 3×10^{-15} m it is attractive, but it becomes repulsive at separations less than about 0.5×10^{-15} m.

This force maintains the nucleons at an approximately constant separation in the nucleus, so nuclear density is about the same for all nuclides, about 2×10^{17} kg m^{-3}.

This compares with the largest density of a macroscopic material, which is about 2×10^4 kg m^{-3}.

Nucleon number and nuclear radius

$\rho = \dfrac{m}{V}$ where ρ is the density

So for a spherical nucleus:

$\rho = \dfrac{A \times m_N}{\frac{4}{3}\pi R^3}$ where m_N is the nucleon mass and R is the nuclear unit radius.

$\therefore R^3 \propto A$ so $R = r_0 A^{\frac{1}{3}}$ where r_0 is a constant, the value for R when A is one (that is, the hydrogen nuclear radius).

Experimental results confirm this relationship between nuclear radius and nucleon number.

Worked example

The table shows data for a range of nuclei.

Element	$r(10^{-15}$ m)	A
C	2.66	12
Si	3.43	28
Fe	4.35	56
Sn	5.49	120
Pb	6.66	208

Use the data to plot a straight-line graph and hence estimate the value of r_0. **(5 marks)**

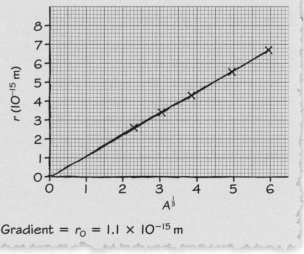

Gradient = r_0 = 1.1×10^{-15} m

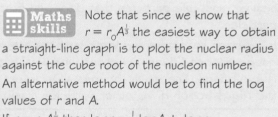

Maths skills Note that since we know that $r = r_0 A^{\frac{1}{3}}$ the easiest way to obtain a straight-line graph is to plot the nuclear radius against the cube root of the nucleon number.

An alternative method would be to find the log values of r and A.

If $r = r_0 A^{\frac{1}{3}}$ then $\log r = \frac{1}{3}\log A + \log r_0$

So a graph of $\log r$ against $\log A$ would yield a straight line with y-intercept equal to $\log r_0$.

A log–log plot can be used to reduce a power relationship to a linear relationship.

Now try this

It is known from electron diffraction experiments that the radius of a gold nucleus (Au-197) is about 7.0×10^{-15} m.

(a) Calculate the density of matter in the nucleus. **(4 marks)**

(b) Determine the radius of an iron nucleus, Fe-56. **(3 marks)**

The Standard Model

All matter is made of elementary particles that occur in two types, called quarks and leptons.

Quarks

There are six **quark** 'flavours', paired in three 'generations'.

1st generation	up	u	down	d
2nd generation	strange	s	charm	c
3rd generation	top	t	bottom	b

Each quark has an antiquark equivalent with the same mass and opposite charge but made of antimatter, so there are 12 possible building blocks in the quark family.

Quarks have fractional charge (a fraction of e), but do not occur singly. They are always found in combination with other quarks or antiquarks.

Particles that are composed of quarks are referred to as **hadrons**. Hadrons interact mostly via the strong interaction.

Leptons

There are six **leptons**, which are paired in three 'generations'.

1st generation	electron	e	electron neutrino	ν_e
2nd generation	muon	μ	muon neutrino	ν_μ
3rd generation	tau	τ	tau neutrino	ν_τ

Each lepton has an antilepton equivalent with the same mass (if any) and opposite charge (if any) but made of antimatter, so there are 12 particles in the lepton family. The positron (antielectron), for example, has the same mass as an electron but a charge of +e.

The electron, muon, and tau all have charge and mass. Neutrinos are electrically neutral and have very little mass.

Leptons interact via the weak interaction, or **weak nuclear force**.

The particle family tree

The lightest and most stable particles in each family make up the first generation. All stable matter is made from particles from the first generation.

The second and third generations in each family consist of heavier, unstable particles that decay to first-generation particles.

In addition to quarks and leptons there are particles called gauge bosons (gluons, photons, and weakons) as well as the Higgs boson. The gauge bosons allow interactions to occur; the Higgs boson gives other particles mass.

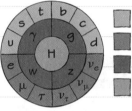

■ quarks
■ leptons
■ gauge bosons
■ Higgs boson

All the particles in the Standard Model. You do not need to be able to reproduce this.

Compare and contrast the properties of hadrons and leptons. **(2 marks)**

Hadrons are composed of quarks, whereas leptons are elementary particles.

Hadrons experience the strong interaction, whereas leptons experience the weak interaction.

The results of particle accelerator experiments suggest that fundamental particles that make up matter are of two types: quarks and leptons. Outline the key features of these two particle types. **(6 marks)**

The quark model of hadrons

Hadrons – made of quarks – can be classified into baryons and mesons.

Quark characteristics

	Charge (e)	Mass (u)		Charge (e)	Mass (u)
u	$+\frac{2}{3}$	0.0025	d	$-\frac{1}{3}$	0.005
c	$+\frac{2}{3}$	1.4	s	$-\frac{1}{3}$	0.1
t	$+\frac{2}{3}$	190	b	$-\frac{1}{3}$	4.5

Each quark has an antiquark with the same mass but opposite charge. The antiquark of u, for example, is the anti-up, written $\bar{u}$, and has charge $-\frac{2e}{3}$.

Properties of hadrons

Hadrons are either uncharged or have a charge that is a whole-number multiple of e.

Baryons are hadrons that are formed when quarks combine together in threes.

Mesons are hadrons that are formed when a quark–antiquark pair combines.

Hadrons experience all four fundamental forces:
- electromagnetism
- gravity
- weak nuclear force
- strong nuclear force.

This is in contrast to leptons, which do not experience the strong interaction.

Baryons

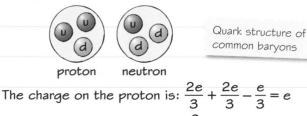

Quark structure of common baryons

proton   neutron

The charge on the proton is: $\frac{2e}{3} + \frac{2e}{3} - \frac{e}{3} = e$

The charge on the neutron is: $\frac{2e}{3} - \frac{e}{3} - \frac{e}{3} = 0$

Antiprotons have a quark structure $\bar{u}\,\bar{u}\,\bar{d}$ so their charge is $-e$. Similarly, antineutrons are made up of $\bar{u}\,d\,\bar{d}$ and thus have no charge.

Mesons

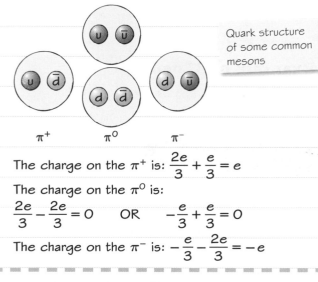

Quark structure of some common mesons

π^+ π^0 π^-

The charge on the π^+ is: $\frac{2e}{3} + \frac{e}{3} = e$

The charge on the π^0 is:
$\frac{2e}{3} - \frac{2e}{3} = 0$ OR $-\frac{e}{3} + \frac{e}{3} = 0$

The charge on the π^- is: $-\frac{e}{3} - \frac{2e}{3} = -e$

Worked example

Determine the charge of particles with a quark structure: (a) u u u; (b) u d s. **(2 marks)**

(a) u u u has charge $\frac{2e}{3} + \frac{2e}{3} + \frac{2e}{3} = 2e$

(b) u d s has charge $\frac{2e}{3} - \frac{e}{3} - \frac{e}{3} = 0$

Worked example

Kaons are strange mesons. Determine the structure of a K^+ meson. **(3 marks)**

It is a meson, so must be a quark–antiquark pair; it is a strange particle, so must include a strange quark.

$u\bar{s}$ since $u\bar{s}$ has charge $\frac{2e}{3} + \frac{e}{3} = e$

Now try this

Λ (lambda) particles are composed of an up quark, a down quark, and another quark. The Λ^0 is neutral. State what type of hadron the Λ^0 is and suggest its quark structure. **(2 marks)**

β⁻ decay and β⁺ decay in the quark model

Radioactive beta decays are governed by the weak nuclear force.

β⁻ decay

The proton number of the nucleus increases while the mass number stays the same. An electron and an antineutrino are emitted.

$$^{A}_{Z}X \rightarrow \,^{A}_{Z+1}Y + \,^{0}_{-1}\beta^- + \,^{0}_{0}\overline{\nu}$$

Within the nucleus:

$$^{1}_{0}n \rightarrow \,^{1}_{1}p + \,^{0}_{-1}\beta^- + \,^{0}_{0}\overline{\nu}$$

The charges (lower row of numbers) are balanced.

In terms of quarks:

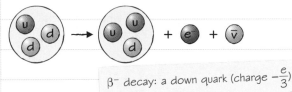

β⁻ decay: a down quark (charge $-\frac{e}{3}$) changes into an up quark (charge $+\frac{2e}{3}$).

β⁺ decay

The proton number of the nucleus decreases while the mass number stays the same. A positron and a neutrino are emitted.

$$^{A}_{Z}X \rightarrow \,^{A}_{Z-1}Y + \,^{0}_{1}\beta^+ + \,^{0}_{0}\nu$$

Within the nucleus:

$$^{1}_{1}p \rightarrow \,^{1}_{0}n \rightarrow \,^{0}_{1}\beta^+ + \,^{0}_{0}\nu$$

In terms of quarks:

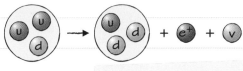

β⁺ decay: an up quark (charge $+\frac{2e}{3}$) changes into a down quark (charge $-\frac{e}{3}$).

Example of β⁻ decay

$$^{14}_{6}C \rightarrow \,^{14}_{7}N + \,^{0}_{-1}\beta^- + \,^{0}_{0}\overline{\nu}$$

Example of β⁺ decay

$$^{18}_{9}F \rightarrow \,^{18}_{8}O + \,^{0}_{1}\beta^+ + \,^{0}_{0}\nu$$

Worked example

Free neutrons are unstable and most decay into protons within a few minutes.

(a) Write a nuclear equation to represent neutron decay. **(2 marks)**

$$^{1}_{0}n \rightarrow \,^{1}_{1}p + \,^{0}_{-1}\beta^- + \,^{0}_{0}\overline{\nu}$$

(b) Write the change in terms of fundamental particles. **(2 marks)**

$$d \rightarrow u + e + \overline{\nu}$$

(c) Suggest how you know that this interaction is governed by the weak nuclear force. **(2 marks)**

The interaction involves leptons - an electron and an antineutrino. These are not affected by the strong nuclear force.

Now try this

Suggest why β⁻ decay occurs for some isotopes and β⁺ decay for others, although the majority of isotopes are neither β⁻ nor β⁺ emitters. **(3 marks)**

A level Module 6

Radioactivity

Some nuclei are unstable and decay by emitting α, β or γ radiation.

Ionising and penetrating power

α, β, and γ radiations are all examples of ionising radiation. When they interact with matter they strip electrons from atoms, freeing electrons and producing positive ions.

- α radiation is highly ionising, therefore it interacts most with matter and is the least penetrating radiation.
- β radiation is moderately ionising, therefore it is more penetrating than α radiation.
- γ radiation is the least ionising, therefore it is the most penetrating.

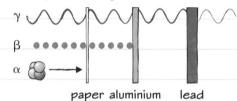

paper aluminium lead

A few centimetres of air will also stop α particles; β particles can travel through some 5 m of air before they are all absorbed. Concrete, like lead, will absorb a lot of γ rays.

Note that γ radiation is never completely absorbed, but its intensity can be reduced considerably by thick lead or concrete.

Investigating the absorption of α, β and γ radiations

The penetrating power of α, β and γ radiation is investigated using a Geiger–Müller tube and counter to detect the radiation passing through different materials placed between the tube and the radioactive source.

The **background radiation** count rate (the number of ionising events per second in the absence of a radioactive source) must first be measured and then subtracted from all recorded count rates. This background count is due to natural radioactivity in rocks and air, as well as cosmic radiation and a small amount caused by human activity.

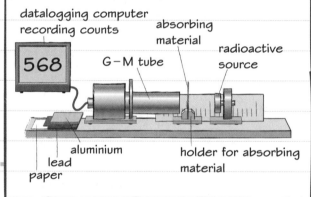

datalogging computer recording counts

absorbing material

radioactive source

G–M tube

aluminium
lead
paper

holder for absorbing material

Worked example

Explain how an electric field and a Geiger–Müller tube could be used to identify which of the three types of radiation were being emitted by a radioactive source. **(4 marks)**

The source of radiation would need to be in an evacuated enclosure. α particles would experience a force in the direction of the electric field, whereas β particles would experience a force in the opposite direction to the field. The γ radiation would not experience a force due to the electric field. Hence the GM tube would detect γ radiation without deflection, and opposite deflections for α and β radiations.

Radiation types

- Alpha (α) radiation consists of helium nuclei.
- Beta (β) radiation consists of fast-moving electrons.
- Gamma (γ) radiation consists of streams of high-energy photons. Hence gamma radiation is short-wavelength EM radiation, typically with a wavelength $< 10^{-11}$ m.

	Charge (e)	Mass (u)
α	2	4
β	–1	0.5×10^{-3}
γ	0	0

Now try this

Radioactive sources are sometimes used as tracers to investigate blood flow in medical diagnosis and fluid flow in pipe systems. Explain why a β source would be suitable for investigating blood flow around the body but unsuitable for water flow around a copper pipe system. **(3 marks)**

Balancing nuclear transformation equations

The emission of ionising radiations from the nucleus causes changes in the nuclear structure.

Alpha emission

When a nucleus emits an α particle – a helium nucleus – it loses two protons and two neutrons. Hence the proton number decreases by 2 and the nucleon number decreases by 4.

$$_Z^A X \to _{Z-2}^{A-4} V + _2^4 \alpha$$

$$_{88}^{226} Ra \to _{86}^{222} Rn + _2^4 \alpha$$

Note that most alpha emitters are massive nuclides whose nuclei are just too big to be stable. The helium nucleus is a particularly stable nuclear configuration and its emission enables the nucleus to lose mass and energy.

Beta⁺ emission

β^+ particles are positrons, and are emitted when a proton turns into a neutron in the nucleus (see page 144).

So when a nucleus emits a β^+ particle the proton number decreases by 1 and the nucleon number stays the same.

$$_Z^A X \to _{Z-1}^A Y + _1^0 \beta^+ + _0^0 \nu$$

$$_{12}^{23} Mg \to _{11}^{23} Na + _1^0 \beta^+ + _0^0 \nu$$

Note that β^+ emitters tend to be nuclides in which there are too few neutrons in the nucleus for stability. β^+ emission increases the neutron-to-proton ratio in the nucleus and hence helps to achieve stability.

Beta⁻ emission

β^- particles are electrons, and are emitted when a neutron turns into a proton in the nucleus (see page 144).

So when a nucleus emits a β^- particle the proton number increases by 1 and the nucleon number stays the same.

$$_Z^A X \to _{Z+1}^A Y + _{-1}^0 \beta^- + _0^0 \bar{\nu}$$

$$_6^{14} C \to _7^{14} N + _{-1}^0 \beta^- + _0^0 \bar{\nu}$$

Note that β^- emitters tend to be nuclides in which there are too many neutrons in the nucleus for stability. β^- emission reduces the neutron-to-proton ratio in the nucleus and hence helps to achieve stability.

Gamma emission

γ radiation is short-wavelength electromagnetic radiation. γ-ray photons are emitted when the nucleus is dropping to its ground state following the emission of an α or β particle. Hence there is no change in either the proton or the nucleon number.

$$_Z^A X \to _Z^A X + _0^0 \gamma$$

$$_{56}^{137} Ba \to _{56}^{137} Ba + _0^0 \gamma$$

Sometimes the γ-ray photons are emitted very shortly after the α or β decay, as in the case of radium (Ra). Sometimes there is a long delay between the decays, as in technetium (Tc).

Complete the nuclear equations:

(a) $_6^7 C \to _\_^{11} B + _\_^\_ \beta^+ + _0^0 \nu$　　**(2 marks)**

$$_6^{11} C \to _5^{11} B + _1^0 \beta^+ + _0^0 \nu$$

(b) $_\_^{225} Ac \to _{87}^\_ Fr + _\_^\_ \alpha$　　**(2 marks)**

$$_{89}^{225} Ac \to _{87}^{221} Fr + _2^4 \alpha$$

(c) $_\_^{212} Bi \to _{84}^\_ Po + _{-1}^0 \beta^- + _0^0 \bar{\nu}$　　**(2 marks)**

$$_{83}^{212} Bi \to _{84}^{212} Po + _{-1}^0 \beta^- + _0^0 \bar{\nu}$$

Note that the sums of the of the nucleon numbers (top line) are the same before and after the decay, as are the sums of the proton numbers (bottom line).

$_{93}^{237} Np$ decays to $_{83}^{209} Bi$ via a series of α and β^- decays. Determine the number of α and the number of β^- decays that take place in this series as a nucleus of Np decays to form a nucleus of Bi.　　**(4 marks)**

Only α decay changes the nucleon number.

$\Delta A = 237 - 209 = 28 = 4 \times 7$

So there must be 7 α particles emitted.

7 α particles would reduce Z by 14, but Z only reduces by 10. There must be 4 β^- decays.

Now try this

$_{92}^{235} U$ is the start of a decay series. In this series there are 7 α decays and 4 β^- decays. Determine the final stable nuclide.　　**(4 marks)**

Radioactive decay 1

Radioactive decay is a spontaneous and random process.

Probabilities

The decay of an unstable nucleus happens spontaneously. The decay is not influenced by temperature or any other external factors.

It is also a random process, as we cannot predict when an individual nucleus will decay.

However, a pattern emerges if the number of unstable nuclei in the sample is large.

The **probability** of a given nucleus decaying in the next unit time interval is fixed for a given nuclide.

The decay constant

The probability of a nucleus decaying in the next unit time interval is called the **decay constant**, λ. λ also equals the fraction of the unstable nuclei decaying in the next unit time interval.

With N undecayed nuclei in the sample, the rate of decay, or activity A, of the sample is given by:

$A = -\dfrac{\Delta N}{\Delta t} = \lambda N$ A has unit s^{-1} or becquerel (Bq).

The minus sign indicates that the number of unstable nuclei in the sample is decreasing.

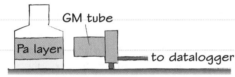 **Practical skills** **Constant-ratio variation**

The constant-ratio variation can be tested experimentally by measuring the activity of an isotope such as protactinium (^{234}Pa).

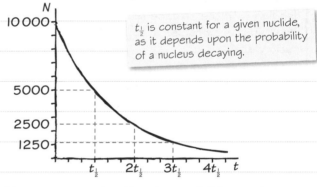

GM tube

Pa layer

to datalogger

^{234}Pa has a half-life of a little over a minute. It is produced in an aqueous layer in a sealed plastic bottle, and the bottle is shaken to transfer Pa to an upper organic layer.

When the layers settle, the GM tube is placed alongside the top layer, and the datalogger used to record the count rate for about 10 min.

The half-life can be determined from the count rate graph produced by the datalogger. The background count rate must be subtracted first.

Half-life

The time taken for the number of unstable nuclei to decrease to a given fraction of its initial value is always the same.

The time taken for the number to decrease to half of its initial value is referred to as the **half-life** of the sample, $t_{\frac{1}{2}}$.

N
10 000
5000
2500
1250
$t_{\frac{1}{2}}$ $2t_{\frac{1}{2}}$ $3t_{\frac{1}{2}}$ $4t_{\frac{1}{2}}$ t

> $t_{\frac{1}{2}}$ is constant for a given nuclide, as it depends upon the probability of a nucleus decaying.

You may recognise the shape of this graph as constant-ratio variation (as in capacitor discharge, see page 123).

The half-life is related to the decay constant by the equation: $\lambda t_{\frac{1}{2}} = \ln 2$

 **Practical skills**   If no datalogger is available, then the count rate can be determined manually.

To reduce uncertainty due to the random nature of the decay, the count should be taken for 10 s at 1 min intervals.

The count with no source present should be taken for 5 minutes twice to obtain an average background count.

Before drawing the graph, all count rates should be corrected by subtracting the background count rate.

Worked example

The half-life of ^{32}P is 14.3 days. Calculate the probability of a given ^{32}P nucleus decaying in the next second.   **(3 marks)**

$t_{\frac{1}{2}} = 14.3 \times 24 \times 3600 = 1.24 \times 10^6$ s

$\lambda t_{\frac{1}{2}} = \ln 2$

$\lambda = \dfrac{\ln 2}{1.24 \times 10^6} = 5.6 \times 10^{-7} s^{-1}$

So the probability of a nucleus of ^{32}P decaying in the next second is 5.6×10^{-7} : 1

Now try this

Radioactive decay is often said to be spontaneous and random. Explain what is meant by 'random' and 'spontaneous', and hence distinguish between these two characteristics of radioactive decay.   **(3 marks)**

Radioactive decay 2

Radioactive decay obeys a constant ratio rule.

Constant ratio

The number of undecayed nuclei obeys a constant-ratio rule:

$$N_0 \rightarrow \frac{N_0}{2} \rightarrow \frac{N_0}{4} \rightarrow \frac{N_0}{8} \rightarrow \ldots$$

$$t = 0 \qquad t_{\frac{1}{2}} \qquad 2t_{\frac{1}{2}} \qquad 3t_{\frac{1}{2}} \qquad \ldots \quad nt_{\frac{1}{2}}$$

$$N_0 \rightarrow \frac{N_0}{2^1} \rightarrow \frac{N_0}{2^2} \rightarrow \frac{N_0}{2^3} \rightarrow \ldots \rightarrow \frac{N_0}{2^n}$$

After n half-lives have elapsed the number of unstable nuclei remaining in the sample is $\frac{1}{2^n} \times$ the initial number of unstable nuclei.

This rule is also valid for time intervals that are not complete half-lives.

Exponential expression

Constant-ratio variations are common in nature.

The mathematical description of such variations is the exponential function.

For radioactive decay we had:

$$-\frac{\Delta N}{\Delta t} = \lambda N$$

This equation can be solved to give an expression for the number of undecayed nuclei N in the sample after a time t.

$$N = N_0 e^{-\lambda t}$$

N_0 is the number of undecayed nuclei at $t = 0$, λ is the decay constant.

🧪 Practical skills — Demonstrating constant-ratio decay

A die is a cube that has a one in six chance of landing with the six face up when it is thrown. You can therefore use it to model constant-ratio decay.

Take a large number of dice and throw them repeatedly. Each time, remove any that have landed showing a six. Note how many dice remain after each throw. If you plot a graph of your results you will find that it models exponential decay; you can calculate the decay constant and half-life.

Spreadsheet modelling

The equation $-\frac{\Delta N}{\Delta t} = \lambda N$ can be used as the basis of a spreadsheet model: a decrease in the number of undecayed nuclei, ΔN, is subtracted from an initial value of N. Repeated iterations model the change ΔN in small time intervals Δt as $\Delta N = -\lambda N \times \Delta t$ (In this model ΔN must be $\ll N$, so Δt must be short).

A similar method is described for modelling capacitor discharge on page 123.

Straight-line graphs

$$N = N_0 e^{-\lambda t} \text{ and } A = -\frac{\Delta N}{\Delta t} = \lambda N$$

So we can write

$$\frac{\Delta N}{\Delta t} = \left(\frac{\Delta N}{\Delta t}\right)_0 e^{-\lambda t} \quad \text{or} \quad A = A_0 e^{-\lambda t}$$

A_0 is the activity (rate of decay) at $t = 0$.

To reduce an exponential function to a linear relationship we use natural logarithms.

$$N = N_0 e^{-\lambda t} \quad \text{so} \quad \ln N = \ln(N_0 e^{-\lambda t})$$

$$\ln N = \ln N_0 - \lambda t$$

Worked example

^{25}Na is a radionuclide. It decays to a stable isotope with a half-life of 59.2 s. Calculate the time taken for the activity of a sample of ^{25}Na to fall below 10% of its initial value. **(3 marks)**

$$\lambda t_{\frac{1}{2}} = \ln 2 \quad \therefore \lambda = \frac{0.693}{59.2} = 0.0117 \, s^{-1}$$

$$N = N_0 e^{-\lambda t} \quad \therefore 0.1 = e^{-0.0117t}$$

$$\therefore t = \frac{\ln 0.1}{-0.0117} = 196.8 \, s. \text{ It would take } 197 \, s.$$

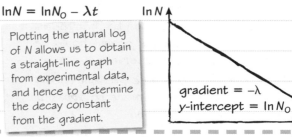

Plotting the natural log of N allows us to obtain a straight-line graph from experimental data, and hence to determine the decay constant from the gradient.

gradient = $-\lambda$
y-intercept = $\ln N_0$

Now try this

Plants take in carbon dioxide from the atmosphere to produce cellulose. A tiny fraction of the carbon atoms in this carbon dioxide are the radioactive isotope ^{14}C. The equilibrium concentration of ^{14}C in living plants produces an activity of 0.267 Bq per g of carbon. When the plant dies, no more ^{14}C is absorbed, and the ^{14}C already present undergoes radioactive decay to ^{14}N. Estimate the age of a piece of wood if 2.0 g of carbon extracted from it has an activity of 0.25 Bq. (The half-life of ^{14}C is 5730 years.) **(4 marks)**

Einstein's mass–energy equation

Einstein's theory of relativity leads to the conclusion that energy and mass are equivalent.

Mass and energy

Einstein discovered that mass and energy are equivalent:

$\Delta E = c^2 \Delta m$, where $c = 3.00 \times 10^8 \, m \, s^{-1}$.

Every reaction that releases energy has an equivalent change in mass.

The law of conservation of energy, must be amended to include **mass–energy** in such situations.

Although mass changes accompanying chemical reactions are too small to measure, those due to nuclear reactions are much larger.

How much energy?

Consider the energy equivalent of a 1 kg mass.

$\Delta E = c^2 \Delta m$, so $\Delta E = (3.00 \times 10^8)^2 \times 1$
$= 9.0 \times 10^{16} \, J$

This is a very large amount of energy. In most nuclear reactions the energy involved is much smaller. In a typical single decay, about 5 MeV of energy is released.

$5 \, MeV = 5 \times 10^6 \times 1.6 \times 10^{-19} = 8 \times 10^{-13} \, J$

$\Delta m = \dfrac{\Delta E}{c^2} = \dfrac{8 \times 10^{-13}}{(3.0 \times 10^8)^2} = 8.89 \times 10^{-30} \, kg$

Mass differences

We can apply Einstein's equation to mass differences observed in radioactive decay, as well as in fission, fusion, annihilation and pair production.

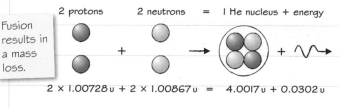

2 protons   2 neutrons   =   1 He nucleus + energy

Fusion results in a mass loss.

$2 \times 1.00728 \, u + 2 \times 1.00867 \, u = 4.0017 \, u + 0.0302 \, u$

1 unified atomic mass unit (u) $= 1.661 \times 10^{-27} \, kg$.

Annihilation and pair production

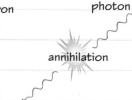

electron   photon

annihilation

positron   photon

Annihilation occurs when a particle and its antiparticle meet. The particles cease to exist and their mass becomes energy. In low-energy collisions, two identical gamma photons are produced.

The reverse process also relies on the principle of mass–energy equivalence. A high-energy photon can produce mass in the form of a particle–antiparticle pair.

Worked example

Complete the equation for ^{11}C decay and calculate the energy released. **(3 marks)**

$^{11}_{6}C \rightarrow \; \ldots B + \; \ldots \beta^+ + \; ^{0}_{0}\nu$

	Mass (u)
Positron	0.0005486
Boron nucleus	11.0061192
Carbon nucleus	11.0076221

$^{11}_{6}C \rightarrow \; ^{11}_{5}B + \; ^{0}_{1}\beta^+ + \; ^{0}_{0}\nu$

$\Delta m = 11.0076221 - 11.0061192 - 0.0005486$

$\Delta m = 9.54 \times 10^{-4} \, u$
$= 9.54 \times 10^{-4} \times 1.661 \times 10^{-27} \, kg$

$\Delta E = (3 \times 10^8)^2 \times 1.58 \times 10^{-30}$
$= 1.43 \times 10^{-13} \, J$

Maths skills It is important to maintain a sufficient number of decimal places when calculating mass differences in nuclear reactions, because the difference in mass is always very small. For example, if we take values to 3 s.f. for proton and neutron masses, we have $m_p = 1.00 \, u$ and $m_N = 1.00 \, u$. Thus the mass difference when 2p and 2n form a He nucleus ($m = 4.00 \, u$) is $4.00 - 2.00 - 2.00 = 0.00$. Using too few decimal places has made the difference in mass disappear!

Note that once the mass difference has been calculated we can revert to a smaller number of s.f. without losing accuracy.

Now try this

1 In a pair-production event, an electron and a positron are created from a gamma photon. Suggest why in this process particles can only be produced in pairs, and why they move outwards in opposite directions. **(2 marks)**

2 When ^{60}Co decays by β^- decay into ^{60}Ni, the decrease in mass is 0.003 u. Calculate the energy released.
$^{60}_{27}Co \rightarrow \; ^{60}_{28}Ni + \; ^{0}_{-1}\beta^- + \; ^{0}_{0}\bar{\nu}$ **(3 marks)**

Binding energy and binding energy per nucleon

The mass defect when a nucleus forms can be interpreted as the binding energy of the nucleus.

Binding energy

When nucleons come together to form a nucleus, energy is released and the mass decreases.

The difference between the mass of the nucleons and the mass of the nucleus is called the **mass defect** of the nucleus, Δm.

The energy equivalent of the mass defect is referred to as the **binding energy** of the nucleus, ΔE. The binding energy is the amount of energy that would have to be supplied to return the nucleus to individual nucleons.

A more useful concept than the total binding energy is the **binding energy per nucleon**, as this is a measure of the stability of the nucleus.

Binding energy per nucleon and nuclear stability

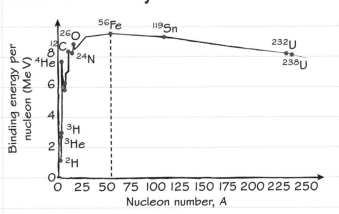

The binding energy per nucleon curve peaks at ^{56}Fe. This is the most stable nuclide.

The binding energy per nucleon for ^{4}He is particularly high for nuclides in this part of the graph. This explains why when heavy unstable nuclei decay they usually emit a He nucleus in α emission.

In general, joining light nuclei together to form larger nuclei results in more stable nuclei and an increase in binding energy, up to a product nucleus of ^{56}Fe. Hence nuclear fusion is a way in which energy can be released.

Similarly, splitting heavy nuclei apart to form lighter product nuclei results in an increase in binding energy per nucleon, down to product nuclei more massive than ^{56}Fe. Hence nuclear fission can result in energy release too.

Calculation of binding energy

To determine the binding energy of a nucleus:
- calculate the mass defect Δm
- use $\Delta E = c^2 \Delta m$ to determine its energy equivalent.

For a nucleus of ^{63}Cu with a mass of 62.91367 u, composed of 29 protons (mass 1.00728 u) and 34 neutrons (mass 1.00867):

$\Delta m = (29 \times 1.00728\,u + 34 \times 1.00867\,u)$
$\qquad - 62.91367\,u$

$\Delta m = 63.50590\,u - 62.91367\,u = 0.59223\,u$

$\Delta m = 0.59223 \times 1.66 \times 10^{-27}\,kg$
$\qquad = 9.831 \times 10^{-28}\,kg$

$\Delta E = (3.00 \times 10^8)^2 \times 9.831 \times 10^{-28}$
$\qquad = 8.848 \times 10^{-11}\,J$

Worked example

Calculate the binding energy per nucleon for ^{56}Fe. **(4 marks)**

	Mass (u)
proton	1.00728
neutron	1.00867
iron nucleus	55.93494

$\Delta m = (26 \times 1.00728\,u + 30 \times 1.00867\,u)$
$\qquad - 55.93494$

$\Delta m = 0.51444\,u = 0.51444 \times 1.661 \times 10^{-27}\,kg$

$\Delta E = (3.00 \times 10^8)^2 \times 8.545 \times 10^{-28}$
$\qquad = 7.69 \times 10^{-11}\,J$

$\Delta E = \dfrac{7.69 \times 10^{-11}}{1.6 \times 10^{-13}} = 481\,MeV$

ΔE per nucleon $= \dfrac{477.6}{56} = 8.59\,MeV$

The S.I. unit of mass, the kg, is large and unwieldy to use for subatomic particles, so you have also been using the atomic mass unit, u. $1\,u = 1.661 \times 10^{-27}\,kg$.

Another unit of mass used by particle physicists is the MeV/c^2 or GeV/c^2. This unit relies upon the Einstein mass–energy equivalence equation. $\Delta m = \Delta E/c^2$, so if we know the energy equivalent of the particle mass, then we can use this value as the particle mass in MeV/c^2.

Now try this

The proton has a mass of 938 MeV/c^2. Calculate the energy available to be converted into photons when a proton annihilates with an antiproton.

(2 marks)

Nuclear fission

Some massive nuclei can undergo fission and split into less massive nuclei, hence releasing energy.

Fission

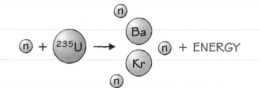

$n + {}^{235}U \rightarrow Ba + Kr + n + n + ENERGY$

Some massive nuclei are **fissile** – they may spontaneously break up into smaller nuclei. The uranium isotope ^{235}U will undergo fission after absorbing a slow neutron. The exact fragment nuclei cannot be predicted, although a number of neutrons are always emitted as part of this process.

As long as at least one of these neutrons is absorbed by another ^{235}U nucleus, a **chain reaction** proceeds in which a considerable amount of energy is released.

Energy release on fission

For nuclei more massive than ^{56}Fe, the binding energy per nucleon decreases as the number of nucleons in the nucleus increases (see page 150).

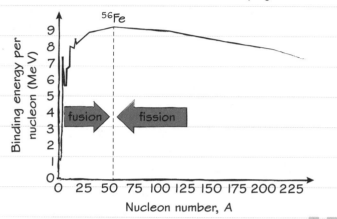

The change in binding energy per nucleon when a ^{235}U nucleus undergoes fission is typically about 0.9 MeV, so releasing about 200 MeV of energy per fission.

Nuclear fission waste

The fragment nuclei formed as a result of the fission in a nuclear reactor are β and γ emitters, and the spent fuel rods contain a very high proportion of α emitter ^{238}U. Nuclear power plants store these in 'spent fuel pools'. The pools are made of thick reinforced concrete with steel liners.

^{238}U can be converted to the fissile element plutonium, which can be used as a fuel in certain nuclear reactors. Other spent fuel may be transported to a reprocessing plant where some radioisotopes may be extracted and then used for medical and other commercial uses.

The rest of the spent fuel can be encapsulated in a radiation-proof canister with a very long expected lifetime and then buried.

To date, no geological repository for spent fuel or other high-level radioactive waste has been built.

Worked example

In a fission of a ^{235}U nucleus, the fission products are ^{144}Ba and ^{89}Kr. Calculate the energy released by one fission. **(4 marks)**

	Binding energy per nucleon (MeV)
Uranium	7.59
Barium	8.41
Krypton	8.71

Binding energies for:

^{235}U: = 235 × 7.59 = 1784 MeV

^{144}Ba: = 144 × 8.41 = 1211 MeV

^{89}Kr: = 89 × 8.71 = 775 MeV

Energy released = 1211 + 775 − 1784
= 202 MeV

Worked example

Explain why the number of neutrons going on to produce further fissions must be exactly equal to 1 in a nuclear fission reactor in a power station. **(3 marks)**

If exactly 1 neutron goes on to produce another fission, the fission rate stays constant.

If less than one neutron goes on to produce another fission, the fission rate will decrease to zero, but if more than one neutron goes on to produce a further fission, the fission rate will increase exponentially.

Now try this

Compare and contrast the key features of radioactive waste disposal for high-level, intermediate-level, and low-level radioactive waste. **(6 marks)**

Nuclear fusion

Light nuclei can undergo fusion and combine to form more massive nuclei, hence releasing energy.

Fusion

For light nuclei less massive than ^{56}Fe, the binding energy per nucleon increases as the number of nucleons in the nucleus increases (see page 150). If the nuclei can come close enough for the strong force to overcome the electrostatic force, fusion can take place. The process releases energy, and is the source of power in stars.

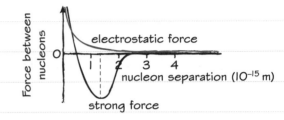

Fusion process

In a star two H nuclei are combined to create ^{4}He and release energy in several steps:

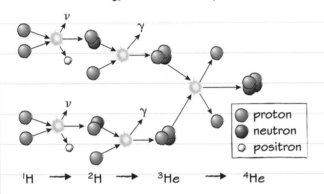

- proton
- neutron
- positron

^{1}H ⟶ ^{2}H ⟶ ^{3}He ⟶ ^{4}He

Extreme conditions are needed for fusion. Only temperatures $> 10^7$ K enable nuclei to overcome electrostatic repulsive forces. Nuclei at very high densities moving at high speeds in such a high temperature **plasma** may then have a collision rate sufficient to maintain fusion.

These conditions are met in the cores of stars, but are hard to achieve in a practical fusion reactor.

Fusion reactors

Practical fusion reactor designs have to resolve the 'containment problem'. The high-temperature plasma required will destroy most materials, and, should the plasma come into contact with a suitable container, it would cool and fusion would stop. Magnetic containment systems which confine the plasma to a toroid (ring doughnut shape), without it touching any field coils, are being tried, although a commercial fusion reactor has not yet been developed.

Fission reactors

In a fission reactor, fuel rods of uranium enriched with fissile ^{235}U are held inside a **moderator**, a material such as water or graphite that will slow the emitted neutrons enough for them to trigger further fission reactions. Control rods can be inserted between the fuel rods or withdrawn as necessary to absorb excess neutrons. The heat generated by the fission reactions is absorbed by a coolant, which in a nuclear power station is used to produce steam to turn generating turbines.

Worked example

Electrostatic repulsion between nucleons must be overcome before fusion is possible. Calculate the magnitude of the electrostatic force between two protons at a separation of 1.3 fm
($e = 1.60 \times 10^{-19}$ C; $\varepsilon_0 = 8.85 \times 10^{-12}$ C^2 N^{-1} m^{-2}).
(2 marks)

$$F = \frac{Qq}{4\pi\varepsilon_0 r^2} = \frac{8.99 \times 10^9 Qq}{r^2}$$

$$= \frac{8.99 \times 10^9 \times (1.6 \times 10^{-19})^2}{(1.3 \times 10^{-15})^2}$$

$$\therefore F = 136 \, \text{N}$$

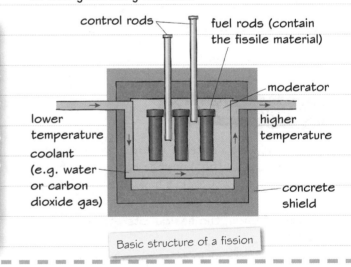

Basic structure of a fission

Now try this

Compare and contrast the advantages and disadvantages of fusion as a potential power source over our current use of fission in nuclear reactors for power.
(6 marks)

Exam skills

This exam-style question uses knowledge and skills you have already revised. Have a look at pages 129 and 134 for a reminder about charged particles in electric and magnetic fields.

Worked example

The electron deflection tube shown here consists of an electron gun inside a vacuum tube. Two parallel metal plates inside the tube can be used to create a variable electric field, and two external coils can be used to create a variable magnetic field. The fields deflect the electron beam.

> You have probably seen a deflection tube in class, but even if you have not you can work out how it works from the description given at the start of the question. Look at the photograph and identify the parts referred to in the question: 1 vacuum tube
> 2 electron gun
> 3 deflector plates
> 4 coils.

(a) Explain how the electrons that form the beam are released from the cathode by thermionic emission. **(2 marks)**

When a metal is heated, the electrons inside it gain kinetic energy, and some are released from the surface of the metal. These can then be accelerated by an applied electric field.

(b) The electrons are then accelerated through a potential difference of 2500 V.

(i) Calculate the kinetic energy of an electron as it enters the region between the deflecting plates.
Give your answer in electronvolts and joules. **(2 marks)**

> You are asked to give the answer in both eV and J. An electronvolt is the energy transferred when a charge e is moved through a p.d. of 1 V. Electrons have charge e and here they are accelerated through a p.d. of 2500 V, so they must each gain 2500 eV of energy.

Kinetic energy = 2500 eV
$$= 2500 \times 1.60 \times 10^{-19} = 4.0 \times 10^{-16} \, J$$

(ii) Calculate the speed of the electron in (i). **(2 marks)**

$$\tfrac{1}{2} mv^2 = 4.0 \times 10^{-16} \, J$$

$$so \; v = \sqrt{\left(\frac{2 \times 4.0 \times 10^{-16}}{9.11 \times 10^{-31}} \right)} = 3.0 \times 10^7 \, m\,s^{-1}$$

> To calculate the speed of the electron in m s⁻¹ you must use kinetic energy in joules. This is because both m s⁻¹ and J are S.I. units (and eV are not).

(c) A separate potential difference is connected across the two deflection plates so that the beam deflects downward as shown. The electric field strength between the plates is 12 000 V m⁻¹.

(i) State the polarity of the lower plate. **(1 mark)**

Positive (it must attract the electrons).

(ii) Calculate the force on an electron when it is between the plates. **(2 marks)**

$$F = Ee = 12\,000 \times 1.6 \times 10^{-19} = 1.9 \times 10^{-15} \, N$$

(d) A student then passes a current through the external coils in order to create a magnetic field through the tube. As the current is increased, the downward deflection decreases.

(i) State the direction of the magnetic field in the tube. **(1 mark)**

> When using Fleming's left-hand rule, remember that the second finger points in the direction of conventional current (+ to −). Here we have a beam of electrons travelling right to left, so they carry **conventional current** from left to right.

Into the page (using Fleming's left-hand rule).

(ii) Calculate the magnetic field strength that would make the beam travel horizontally through the tube (assume that the electric field is unchanged). **(3 marks)**

The magnetic force and electric force on the electron must be equal and opposite:

$$eE = Bev$$

$$B = \frac{E}{v} = \frac{12\,000}{3.0 \times 10^7} = 4.0 \times 10^{-4} \, T$$

> Don't forget the unit.

> Notice how this answer is set out. You might consider calculating a value for the magnetic force and then setting it equal to the value of the electric force from part (c)(ii), to calculate B. This is perfectly acceptable, but the algebraic approach used here is simpler!

Production of X-ray photons

High-energy electrons fired into a heavy metal target produce short-wavelength radiation.

The X-ray tube

Electrons are produced from a heated filament (cathode) in a vacuum and accelerated through a large p.d. towards an anode.

The anode contains a heavy metal target, usually tungsten. When the electrons collide with the target the electron energy is mainly transferred to thermal energy in the target.

A small fraction of the electrons' kinetic energy is transferred to very high frequency (short wavelength) radiation in the X-ray region of the electromagnetic spectrum.

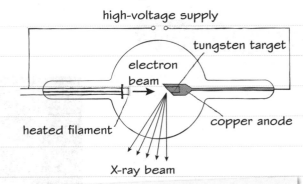

X-rays are released from the angled target when electrons collide with it. The main body of the anode is made from copper so that thermal energy can be conducted away from the target.

The spectrum from an X-ray tube

The X-ray spectrum produced has two main parts: a continuous spectrum characteristic of the accelerating p.d., and a line spectrum characteristic of the target element.

The continuous spectrum is produced as the electrons hit the tungsten target and are rapidly decelerated. The maximum frequency of X-ray photons is determined by the accelerating p.d., V_a.

The characteristic spectrum is produced as high-energy electrons displace electrons from the lowest energy states in the target atoms. Electrons cascade down to fill up these energy states, so high-frequency photons are emitted at frequencies characteristic of the target element.

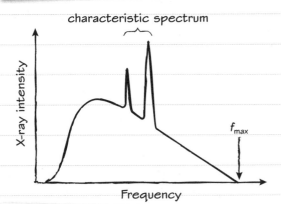

The maximum frequency corresponds to all of an electron's kinetic energy being transferred to a photon.

$$hf_{max} = eV_a$$

Worked example

Calculate the minimum wavelength of radiation emitted from an X-ray tube when the anode potential is 50 kV.   **(3 marks)**

Photon energy = $eV_a = hf_{max}$

$$f_{max} = \frac{eV_a}{h} = \frac{1.61 \times 10^{-19} \times 50 \times 10^3}{6.63 \times 10^{-34}}$$

$$\therefore f_{max} = 1.21 \times 10^{19}\,\text{Hz}$$

$$\lambda_{min} = \frac{c}{f_{max}} = \frac{3 \times 10^8}{1.21 \times 10^{19}} = 2.5 \times 10^{-11}\,\text{m}$$

Worked example

Explain how the spectrum of radiation produced by an X-ray tube changes as the anode voltage is increased.   **(3 marks)**

- The intensity of the radiation increases as the rate of incidence of electrons on the target increases.

- The maximum frequency of radiation produced increases as the kinetic energy of the electrons increases.

- The characteristic lines are dependent upon the target element, and so do not move.

Now try this

Explain why it is important for the anode to be made of a good thermal conductor, and suggest how excess thermal energy could be safely dissipated in the X-ray tube.   **(3 marks)**

X-ray attenuation mechanisms

There are a number of ways in which X-rays interact with matter.

Simple scattering

Simple (Rayleigh) scattering can occur for photon energies <20 keV. This is sometimes referred to as elastic scattering. Low-energy X-ray photons are scattered by electrons in an atom. The photon energy is unchanged in the scattering process.

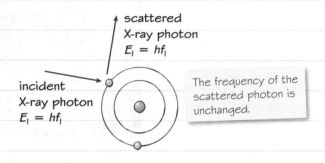

scattered
X-ray photon
$E_1 = hf_1$

incident
X-ray photon
$E_1 = hf_1$

The frequency of the scattered photon is unchanged.

Compton scattering

Compton scattering is an example of inelastic scattering. High-energy X-ray photons are scattered by hitting a free electron. The energy of the scattered photon is less than that of the incident photon, as energy is transferred to the electron.

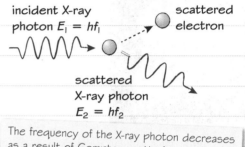

incident X-ray
photon $E_1 = hf_1$

scattered
electron

scattered
X-ray photon
$E_2 = hf_2$

The frequency of the X-ray photon decreases as a result of Compton scattering.

Pair production

When the X-ray photon energy is greater than 1 MeV, the photon can interact with the nucleus of an atom, producing an electron–positron pair. This effect is called **pair production**, and it dominates at photon energies above 5 MeV.

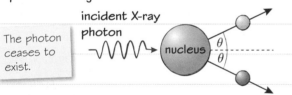

incident X-ray
photon

The photon
ceases to
exist.

nucleus

θ
θ

The positron goes on to annihilate with another electron to produce two photons (see page 149).

Photoelectric effect

At higher energies an X-ray photon may make an inelastic collision with an electron in an atom. All of the photon energy is transferred to the electron, so the photon ceases to exist. The electron is ejected from the atom, and the effect is an example of the **photoelectric effect** (see page 76).

The X-ray photons are absorbed. The attenuation of the X-ray beam with distance in the material is exponential.

$I = I_0 e^{-\mu x}$, where μ is the **attenuation** (or absorption) **coefficient** and x is the absorber thickness. μ depends upon the photon energy.

Worked example

Calculate the minimum photon frequency for pair production to be able to occur ($m_e = 9.1 \times 10^{-31}$ kg; $h = 6.63 \times 10^{-34}$ J s; $c = 3.00 \times 10^8$ m s^{-1}). **(3 marks)**

$\Delta E = c^2 \Delta m = (3.0 \times 10^8)^2 \times 2 \times 9.1 \times 10^{-31}$

$\therefore \Delta E = 1.638 \times 10^{-13}$ J

$\therefore f = \dfrac{1.638 \times 10^{-13}}{6.63 \times 10^{-34}} = 2.5 \times 10^{20}$ Hz

Worked example

The intensity of an X-ray beam is reduced to 17.5% of its initial value after passing through 2.5 cm of aluminium. Calculate a value for the absorption coefficient μ. **(2 marks)**

$I = I_0 e^{-\mu x}$

$0.175 = e^{-0.025\mu}$

$\therefore \mu = \dfrac{\ln(0.175)}{-0.025} = 70 \text{ m}^{-1}$

Now try this

1 When considering wave–particle duality, Compton scattering is sometimes quoted as evidence for the particle nature of electromagnetic radiation. Suggest why a particle model is more appropriate than a wave model when describing the Compton effect. **(3 marks)**

2 An X-ray beam is attenuated to 13% of its initial value after passing through a 12 mm thickness of metal plate. Calculate the absorption coefficient μ. **(3 marks)**

X-ray imaging and CAT scanning

A variety of techniques can be used to improve the quality of X-ray images of the body.

Standard X-ray photographs

Different tissues attenuate an X-ray beam by different amounts. If there is a marked variance between the densities of two organs, natural contrast enables the structures to be seen.

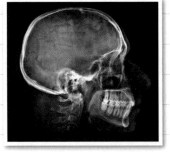

X-rays are often used to obtain images of bones, the densest organs in the body, but they can be used for other organs too.

If two organs have similar densities, it is not possible to distinguish them. The density of a hollow organ can be changed by filling it with a **contrast medium**.

Contrast media – barium

X-ray images of soft tissues can be improved by including a **contrast medium**. For example, the colon and rectum can be imaged with X-rays once a liquid containing barium salts is put into the rectum. Barium has a high atomic number and absorbs X-rays strongly.

Air can be introduced into the rectum and colon to further enhance the X-ray image.

CAT scanning

Computerised axial tomography (CAT) scanning provides detailed images of inside the body.

The CAT scanner has an X-ray tube housed in a ring. It sends a thin fan-shaped beam of X-rays through the patient's body towards detectors on the opposite side of the ring. The X-ray tube and detectors rotate around the patient, recording many images from different angles. These are assembled by software to create cross-sections through the body that can be viewed on screen.

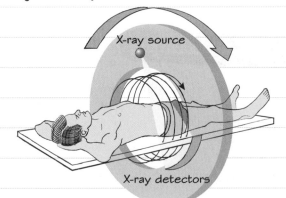

X-ray source

X-ray detectors

The advantages of CAT scans over X-rays include the ability to process the data to produce a variety of views, and to selectively enhance or remove structures from the images.

Contrast media – iodine

The average atomic number of structures such as blood vessels can be increased by filling them with liquid of higher atomic number; one such medium is iodine.

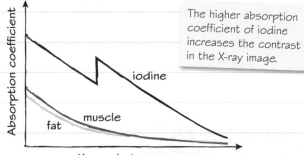

The higher absorption coefficient of iodine increases the contrast in the X-ray image.

Worked example

In 2014, the BBC reported on a rise in the number of CAT scans being performed, exposing people to potential health risks. Explain why CAT scans might be a health risk. **(2 marks)**

X-rays are a form of ionising radiation; therefore they can cause changes in the DNA structure of cells. CAT scans use higher X-ray doses than normal X-ray imaging.

Now try this

Suggest why CAT scanning techniques have largely superseded X-ray imaging techniques using barium salts for the lower digestive tract . **(2 marks)**

Medical tracers

Radioisotopes are used in diagnostic medicine to track processes inside the patient.

Technetium-99m

^{99m}Tc is used as a **radioactive tracer** in medicine. It emits gamma rays with a wavelength about the same as X-rays used in conventional diagnostic equipment, and can be detected outside the body by a gamma camera.

The half-life for gamma emission from the ^{99m}Tc radioisotope is 6.0 hours, long enough for scanning procedures but short enough to keep radiation exposure low for the patient.

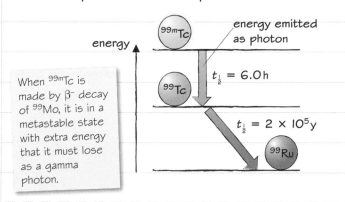

When ^{99m}Tc is made by β^- decay of ^{99}Mo, it is in a metastable state with extra energy that it must lose as a gamma photon.

Use of ^{99m}Tc

^{99m}Tc is used for medical and research purposes, such as evaluating the condition of various organs and in blood flow studies.

Nuclear cardiology is a non-invasive procedure used to show the heart chambers. Tc is injected into a vein, attaches itself to red blood cells and circulates around the body.

A gamma camera traces the Tc as it moves through the heart. A computer then processes the images to make it appear as if the heart is moving.

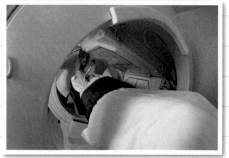

Typical images from a heart stress test using technetium scanning.

Fluorine-18

^{18}F is a radioisotope that decays by positron emission with a half-life of 110 minutes. The short half-life makes it particularly suitable for medical tracer applications.

^{18}F is usually used in PET (positron emission tomography) scans.

Because of its short half-life it has to be made *in situ*, using either a cyclotron or a linear particle accelerator.

Production and use of ^{18}F

A target of water molecules formed with oxygen-18 is bombarded with protons to make ^{18}F. The protons must be accelerated to very high speed in order to overcome the repulsion of the positively charged target nuclei.

^{18}F can be used to make radiolabelled glucose, which is preferentially taken up by active cancer cells, making them visible in scans.

Worked example

A sample of fluorine-18 is prepared with an initial activity of 340 MBq. Calculate the activity of the sample 4 hours later. ($t_{\frac{1}{2}}$ = 110 minutes.)   **(2 marks)**

$$\text{number of half lives} = \frac{4 \times 60}{110} = 2.18$$

$$A = \frac{340}{2^{2.18}} = 75 \text{ MBq}$$

Worked example

Complete the nuclear equation for the production of fluorine-18 from oxygen-18:

$$^{-}_{8}O + ^{-}_{-}p \rightarrow ^{-}_{-}F + ^{-}_{-}n \qquad \textbf{(2 marks)}$$

$$^{18}_{8}O + ^{1}_{1}p \rightarrow ^{18}_{9}F + ^{1}_{0}n$$

Now try this

A tracer technique is being used as a diagnostic procedure in a hospital.
Explain why a γ emitter might be preferred to a β emitter.   **(2 marks)**

The gamma camera and diagnosis

Gamma cameras image the radiation from a tracer introduced into the patient's body.

Gamma cameras

A low-mass dose of a radioisotope such as ^{99m}Tc is injected into the bloodstream. Radioactive decay produces gamma photons, which are emitted from the patient and detected by a gamma camera.

The gamma camera consists of a collimator, a **scintillator**, **photomultiplier** tubes, circuitry and a computer to produce a display image.

The photons released by ^{99m}Tc are ideal for imaging, as they penetrate living tissue but not the sides of the gamma camera and collimator.

The radiation risk involved in gamma camera analysis is very low compared with the potential benefits, because only a small amount of radioisotope is injected and so the procedure results in relatively low radiation exposure to the patient.

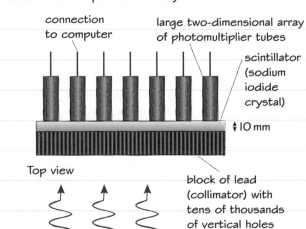

Gamma camera techniques

Gamma cameras image the processes of the body, whereas X-ray techniques image the structure of the body.

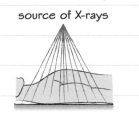

source of X-rays   The γ-ray photons are produced inside of the patient.

X-ray detector plate   gamma camera

The patient lies on a table, which slides between camera heads suspended over the table. Most photons are scattered in the body or unable to penetrate the collimator, so it may take up to 15 min to generate an acceptable image.

The collimator only allows head-on photons to reach the scintillator, which absorbs γ-ray photons and emits visible light. The photomultiplier tubes produce a current pulse when a flash is detected from the scintillator. The current pulses from the array of photomultiplier tubes allow a real-time image to be produced that mirrors the emission of gamma photons from the patient.

Photomultipliers

A photomultiplier tube consists of an earthed cathode that emits electrons when photons are incident upon it via the photoelectric effect

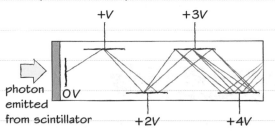

The photoelectrons are accelerated towards an anode (or dynode), where each electron releases at least 2 electrons. These are accelerated to a second dynode where the same process occurs. Hence each dynode increases the number of electrons, so each incident photon causes a pulse of current.

Worked example

A photomultiplier tube has 10 dynodes. Each electron incident upon a dynode emits a further two electrons. Calculate the charge in the pulse resulting from a photon releasing three electrons from the cathode ($e = 1.60 \times 10^{-19}$C). **(2 marks)**

$N = 3 \times (2)^{10} = 3072$

$Q = 3072 \times 1.6 \times 10^{-19} = 4.9 \times 10^{-16}$ C

Now try this

Explain how images produced from a gamma camera differ from X-ray images, and how this enables different diagnostic information to be gained.

(2 marks)

PET scanning and diagnosis

Positron emission tomography, or PET, is used to image organs and tissues in the body to show how they are working.

Positron emission tomography

A radioactive isotope such as fluorine-18 that decays by emitting positrons is introduced into the patient. When a positron meets an electron inside the patient, annihilation (see page 149) produces two gamma rays of equal energy which travel in opposite directions.

PET scanning

The PET scanner detects these gamma rays to produce a high-resolution image of organs and tissues in the body and to show how they are working.

Positron-emitting isotopes of biologically active elements are used in PET scanning procedures. Fluorine-18 (see page 157), can be incorporated into a radioactive substance similar to glucose. This is taken up by cancer cells, making it possible to locate cancerous tumours, and by the brain, so PET scanning can also be used to map brain function. This can lead to the diagnosis of conditions such as Alzheimer's disease.

Tomography

The word 'tomography' comes from the Greek words 'tomos' meaning 'section', 'slice', and 'graphein' meaning 'to draw'. Tomography creates a cross-sectional view (or slice) of the object which is being investigated. The process synthesises detected signals from different directions, allowing images to be formed of the inside of the patient's body without cutting the patient open. These can be joined together by computer technology to produce 3D images.

CAT scanning, another form of tomography, is covered on page 156.

The computer can match the pairs of gamma photons detected and use them to locate the tissue in which the tracer isotope has collected.

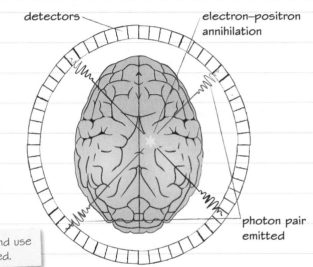

detectors

electron–positron annihilation

photon pair emitted

Worked example

Carbon-11 can be used in PET scans. It can be produced by firing high-speed protons at nitrogen atoms. The nuclear reaction for the production of carbon-11 is shown below. Complete the equation and deduce the particle X emitted in this process.

$^{14}_{7}\text{N} + {}^{...}_{...}\text{p} \rightarrow {}^{11}_{6}\text{C} + {}^{...}_{...}\text{X}$ **(2 marks)**

$^{14}_{7}\text{N} + {}^{1}_{1}\text{p} \rightarrow {}^{11}_{6}\text{C} + {}^{4}_{2}\text{He}$

X is an alpha particle.

Worked example

A positron with energy 0.63 MeV annihilates with a stationary electron. Calculate the energy of the gamma photons produced ($e = 1.60 \times 10^{-19}$ C; $m_e = 9.11 \times 10^{-31}$ kg; $c = 3.00 \times 10^8$ m s^{-1}).   **(2 marks)**

$\Delta E = c^2 \Delta m$

$\therefore \Delta E = (3.00 \times 10^8)^2 \times 2 \times 9.11 \times 10^{-31}$

$\therefore \Delta E = 1.64 \times 10^{-13} \text{J} = \dfrac{1.64 \times 10^{-19}}{1.60 \times 10^{-13}} = 1.03 \text{ MeV}$

So total energy available for photons is 1.66 MeV

So each photon has energy $\dfrac{1.66}{2} = 0.83$ MeV

Now try this

In a PET scanner, a count is recorded when gamma rays are detected in the ring of detectors around the patient. Explain how gamma rays are produced inside the patient and why a count is only detected if two detectors are activated by a gamma ray simultaneously.

(3 marks)

Ultrasound

Ultrasound is a sound wave with a frequency too high for the human ear to detect.

Ultrasound

Ultrasound is any sound wave with a frequency above about 20 kHz. In medical applications, a range of 2–10 MHz is usually used.

Typical medical applications of ultrasound include imaging organs inside the body, measuring blood flow rate, and measuring heart rate.

The ultrasound is produced and detected by a transducer. This uses the **piezoelectric effect** as a mechanism to transfer energy between an electrical circuit and an ultrasound wave.

Piezoelectric crystals

When an alternating p.d. is applied, the piezoelectric crystal vibrates, producing an ultrasound wave

When made to vibrate by an ultrasound sound wave, the piezoelectric crystal produces an alternating e.m.f.

Ultrasound technique

Echolocation is a distance-measuring technique which relies on timing the interval between making a sound and detecting the reflected signal.

A piezoelectric crystal emits an ultrasound pulse that is reflected from a boundary (for example the edge of an internal organ such as a kidney). The reflected pulse returns, causing the crystal to vibrate, and an e.m.f. is produced.

To carry out an ultrasound scan, a coupling gel is used to minimise reflections from the first boundary, the patient's skin.

Calculation

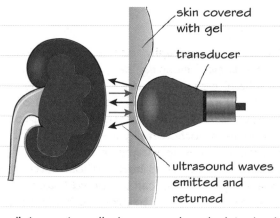

skin covered with gel

transducer

ultrasound waves emitted and returned

The distance travelled one way is calculated using:

Distance travelled to boundary

$$= \frac{\text{wave speed} \times \text{time for return journey}}{2}$$

Worked example

Explain why the ultrasound pulses used in medical imaging must be of short duration.   **(2 marks)**

The transducer acts as emitter and receiver of ultrasound waves. Hence the emitted pulse must stop before the echo arrives back at the transducer, so that the incident and reflected wave pulses do not overlap.

This time is very short, since the distances travelled are very short.

Worked example

Explain why a coupling gel is needed when producing an ultrasound image of an unborn fetus to obtain a high quality image.   **(2 marks)**

A coupling gel is needed between the probe and the mother's skin to exclude air, because the air–skin boundary would reflect as well as transmit the ultrasound.

The gel ensures maximum transmission of the ultrasound waves into the body.

Now try this

State and explain one advantage and one disadvantage of ultrasound compared with X-rays in medical imaging.

(2 marks)

Acoustic impedance and the Doppler effect

When ultrasound waves meet a boundary some energy is reflected and some is transmitted.

Acoustic impedance

When an ultrasound pulse reaches a boundary only some of the wave pulse is transmitted across the boundary. The ratio of the reflected intensity I_r to the incident intensity I_0 is given by the equation:

$$\frac{I_r}{I_0} = \left(\frac{Z_2 - Z_1}{Z_2 + Z_1}\right)^2$$

Z is the **acoustic impedance** of the medium and is equal to the product of the density of the medium and the wave speed in the medium.

$$Z = \rho c$$

Image construction

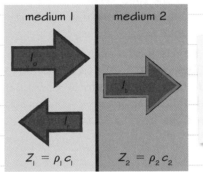

$$Z_1 = \rho_1 c_1 \qquad Z_2 = \rho_2 c_2$$

Multiple reflections can be used to determine distances to two (or more) boundaries and thus can be used to construct an image.

Measuring blood flow rate

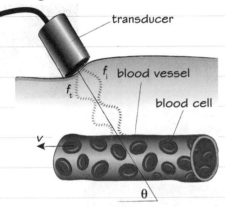

transducer

f_i blood vessel

f_t

blood cell

v

θ

If the transducer is placed above the chest, the frequency change caused by blood rushing away from the heart enables the heart rate to be deduced.

Using the Doppler effect

When an ultrasound transducer is held near a main artery, the blood cells reflect the sound waves.

Waves reflected from cells travelling towards the transducer are detected with a higher frequency than the incident waves. Waves from cells travelling away have a lower frequency. This is the **Doppler effect**.

The Doppler effect is reviewed on page 117.

$$\frac{\Delta f}{f} = 2 \times \frac{v \cos\theta}{c}$$

The factor of 2 arises because the cells act as moving receivers and moving sources, giving a double Doppler shift.

Where v is the speed of the blood cells and the ultrasound transducer is positioned at an angle θ to the movement of the blood cells.

The shift in frequency is directly proportional to the speed of the blood cells, and so measuring the change in frequency allows the blood flow rate to be calculated.

Worked example

Ultrasound is incident on an air–skin boundary. Calculate the percentage of the incident intensity transmitted into the skin ($Z_{air} = 429$ kg m^{-2} s^{-1}; $Z_{skin} = 1.65 \times 10^6$ kg m^{-2} s^{-1}). **(2 marks)**

$$\frac{I_r}{I_0} = \left(\frac{Z_2 - Z_1}{Z_2 + Z_1}\right)^2 \text{ and } \frac{I_t}{I_0} = 1 - \frac{I_r}{I_0}$$

$$\therefore \frac{I_t}{I_0} = 1 - \left(\frac{[1.65 \times 10^6 - 429]}{[1.65 \times 10^6 + 429]}\right)^2 = 1.0 \times 10^{-3}$$

So 0.1% of the incident intensity is transmitted.

🖩 **Maths skills** Watch out for powers-of-ten errors and forgetting to square the bracket.

When carrying out a subtraction of two similar magnitude values, rounding errors can affect the final answer that you obtain. Here the value for $\frac{I_r}{I_0}$ is very close to 1.

Always retain enough significant figures in any intermediate answers that you work out, so that your final answer will be accurate.

Now try this

(a) Use the answer to the worked example to explain the benefit of a coupling gel used between an ultrasound probe and a patient's skin. **(2 marks)**

(b) State and explain what the ideal value of the acoustic impedance would be for such a gel. **(2 marks)**

Ultrasound A-scan and B-scan

Ultrasound scans can be used to determine distances (A-scans) or produce images (B-scans).

A-scans

A-scans or amplitude scans are used to measure distances only. The transducer emits an ultrasound pulse and the time taken for the pulse to reflect from an object and return is used to determine the distance.

A typical A-scan consists of a series of amplitude peaks on an oscilloscope (CRO) screen. Each peak corresponds to a reflection from a boundary. The scans can be difficult to interpret, because of the multiple reflections from various boundaries.

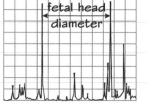

Two common applications of A-scans include measurements in the eye and measurements of fetal skull size.

B-scans

B-scans or brightness scans can be used to make two-dimensional images. The transducer produces multiple ultrasound pulses and, as the transducer is swept across the area, the signals from the reflected pulses are used to produce a series of dots on a screen. The brightness of each dot is determined by the intensity of the reflected pulse.

Successive B scans are made as the transducer probe is rocked sideways on the patient. Each static B scan is added to form a fan-shaped brightness image, which is an image of a cross-section through the body.

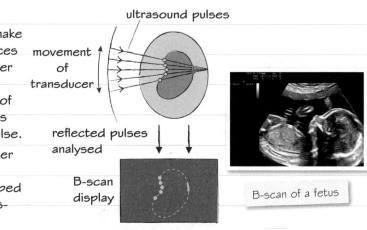

ultrasound pulses

movement of transducer

reflected pulses analysed

B-scan display

B-scan of a fetus

Worked example

The diagram shows an A-scan of an internal organ. The oscilloscope trace shows reflected pulses coming from the front and rear surfaces of the organ. The horizontal scale is 0.02 ms/div.

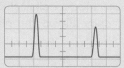

Use the scan to estimate the organ thickness. Ultrasound speed in the organ is $1500\,\mathrm{m\,s^{-1}}$. **(2 marks)**

Time difference = $4 \times 0.02 \times 10^{-3}\,\mathrm{s}$
$$= 8 \times 10^{-5}\,\mathrm{s}$$

Thickness of organ = $\dfrac{1500 \times 8 \times 10^{-5}}{2}$

$$= 0.06\,\mathrm{m}$$

The 2 arises because the ultrasound that creates the second pulse has travelled across the organ and back again.

Worked example

In the A-scan in the first worked example, the height of the second pulse is smaller than that of the first pulse.

Give two reasons why this should be the case. **(2 marks)**

The ultrasound pulse will be attenuated as it passes through the organ.

The ultrasound pulse will then be only partially reflected at the second boundary; some energy will be transmitted onwards out of the organ.

Ultrasound energy is reflected from the first boundary on the return path to the transducer and so the transmitted pulse intensity is reduced.

Now try this

(a) Describe how ultrasound is used to obtain an image of an unborn fetus. **(4 marks)**

(b) Explain why ultrasound is preferable to other methods of imaging the fetus. **(2 marks)**

Answers

AS & A level

1. Quantities and units

1. The derived unit is $N\,m^2\,kg^{-2} = kg\,m\,s^{-2} \times m^2\,kg^{-2}$
   $= kg^{-1}\,m^3\,s^{-2}$

2. $5.0 \times 10^{-4}\,m\,s^{-1}$

3. $6.4 \times 10^{-8}\,m^3$

2. Estimating physical quantities

1. (a) $5.972 \times 10^{24}\,kg$
   (b) $1.673 \times 10^{-27}\,kg$

2. Divide the mass of a human being by the mass of a proton and round off. Order of 10^{28} protons.

3. Mass of elephant is of order $10^3\,kg$ so weight is of order $10^4\,N$. Area of one foot is about $30\,cm \times 30\,cm$ i.e. about $1000\,cm^2 = 10^{-1}\,m^2$. Pressure is about $10^4 \div (4 \times 10^{-1}) = 2.5 \times 10^4\,Pa$ or of the order of $10^4\,Pa$.

3. Experimental measurements

(a) $T = 0.722\,s$

(b) $T = 0.723\,s$

(c) The spread of values shows that there is a random error in the measurement. The effect of this has been reduced by taking an average over all five trials. This means that whilst the actual values in (a) and (b) are similar, the uncertainty associated with the answer in (b) is smaller.

4. Combining errors

The percentage error in s must be added to $2 \times$ the percentage error in t (since t is squared) in order to find the percentage error in g.

$g = 9.80 \pm 0.15\,m\,s^{-2}$

5. Graphs

1. $R = \dfrac{(22.6 + 22.0)}{2} = 22.3\,\Omega$ $\triangle R = \dfrac{(22.6 - 22.0)}{2} = 0.3\,\Omega$ so he should give $R = 22.3 \pm 0.3\,\Omega$

2. (a) $10.0 \pm 0.1\,cm$
   (b) $\frac{2}{51} \times 100 = 4\%$

6. Scalars and vectors

1. (a) Average speed $= \dfrac{distance}{time} = \dfrac{400}{80} = 5.0\,m\,s^{-1}$
   (b) Average velocity $= \dfrac{displacement}{time} = 0\,m\,s^{-1}$

2. I would need to swim at an angle to the west of north so that my component of velocity due west equals the flow velocity of the river due east. The angle to north is θ where $\tan\theta = \frac{1.2}{2.0} = 0.60$ and $\theta = 31°$ west of north.

7. Vector triangles

1. $N = W\cos20° = 12\,000\cos20° = 11\,300\,N$
   $G = W\sin20° = 12\,000\sin20° = 4100\,N$

2. $31°$ west of north.

8. Resolving vectors

1. $3.0 \times 3600 \times \cos65° = 4600\,m$ north.

2. $40\cos32° \times 8.0 = 270\,J$.

9. Describing motion

(a) $10.2\,m\,s^{-1}$

(b) $12.0\,m\,s^{-1}$

(c) $0\,m\,s^{-2}$ because the velocity is constant. The displacement in each second is $12\,m$ so the velocity is constant at $12\,m\,s^{-1}$.

(d) $9.83\,s\,(10 - \frac{2}{12}\,s)$.

10. Graphs of motion

1. (a) It reaches its highest point at $t = 3\,s$. This is where the upward velocity has decreased to zero, so the pellet has stopped rising and is about to start falling.
   (b) This is equal to the area under the graph for the first $3\,s$. Area of triangle $= 0.5 \times 30 \times 3 = 45\,m$.
   (c) The gradient is constant.
   (d) Acceleration is equal to the gradient.
   $a = \dfrac{\triangle v}{\triangle t} = -\dfrac{30}{3} = -10\,m\,s^{-2}$ (downward).

11. SUVAT equations of motion

1. (a) $a = \dfrac{(v - u)}{t} = \dfrac{(50 - 0)}{10} = 5.0\,m\,s^{-2}$
   (b) $s = \frac{1}{2}(u + v)t = \frac{1}{2}(50 + 0) \times 10 = 250\,m$

2. $u = 15.0\,m\,s^{-1}$, $a = 0.50\,m\,s^{-2}$, $v = 25\,m\,s^{-1}$ $s = ?$
   The correct equation is $v^2 = u^2 + 2as$ but this needs to be rearranged to make s the subject of the equation:
   $s = \dfrac{(v^2 - u^2)}{2a} = \dfrac{(25^2 - 15^2)}{1.0} = 400\,m$

12. Acceleration of free fall

1. (a) The fourth result ($476\,ms$). It should be ignored when calculating the average time.
   (b) $497\,ms$ (ignoring anomalous result).
   (c) $g = 9.72\,m\,s^{-2}$

13. Vehicle stopping distances

1. (a) $12.5\,m$
   (b) $31.25\,m$
   (c) $43.75\,m$, assuming constant braking force and the same reaction time.

14. Projectile motion

1. $143\,m$

2. $318\,m$; in practice, frictional forces will cause the cannonball to lose energy so that its time of flight and therefore its range are reduced.

15. Types of force

(a) $W = mg = 2450\,N$

(b) $2450\,N$ upward. The boat has no vertical acceleration so resultant force must be zero.

(c) $a = 0.20\,m\,s^{-2}$

16. Drag

(a) The air in the Earth's atmosphere provides drag but the Moon has no atmosphere so no upward force counteracts the weight of the feather accelerating it downward.

(b) $F = W - D = mg - D = 0.2 \times 9.81 - 1.0 = 0.96\,N$,
   $a = \frac{F}{m} = 4.8\,m\,s^{-2}$

(c) Drag increases with velocity. However large the forward force from the engine, it will eventually be balanced by an equal but opposite force from drag.

17. Centre of mass and centre of gravity

1 As the bottle is tilted, the centre of gravity moves sideways. Eventually it passes over the edge of the bottle in contact with the table. The weight of the bottle then creates a resultant turning effect (moment) in the direction the bottle is being pushed, so the bottle continues to rotate in that direction and topples over.

2 As the bottle is filled with water, the centre of mass falls at first and then rises. It is lowest when the bottle is about $\frac{1}{4}$ full of water.

18. Moments, couples and torques

Moment = $1.2 \cos(30°) \times 60 \times 9.8 = 611$ Nm

19. Equilibrium

(a) 200 N

(b) $L = \dfrac{400 \times 0.6 \times \cos 30°}{1.2} = 173$ N

(c) The vertical component of the lifting force has reduced so the vertical component of P must increase to ensure equilibrium of forces in this direction. In addition to this, there is now a horizontal component of the lifting force directed towards the left. This must be balanced by a horizontal component of the contact force at P acting to the right. The contact force at P is therefore larger and directed at an angle to the right of the vertical.

20. Density and pressure

1 Excess pressure $p = h\rho g = 10 \times 1030 \times 9.81 = 101\,040$ Pa, which is close to the value of atmospheric pressure at the surface (around 10^5 Pa).

2 Pressure $p = h\rho g = 0.760 \times 13\,600 \times 9.81 = 101\,400$ Pa, which is close to 10^5 Pa.

21. Upthrust and Archimedes' principle

1 The weight of the fully submerged object is given by $W = A(h_2 - h_1)\rho_{object}g$. If $\rho_{object} > \rho_{fluid}$ then $W > F_{upthrust}$ so there is a downward resultant force on the object and it sinks. If $\rho_{object} < \rho_{fluid}$ then $F_{upthrust} > W$ so it floats.

2 (a) Pressure due to seawater is
$p = h\rho g = 4.16 \times 1030 \times 9.81 = 42\,000$ Pa.

(b) $F = pA = 42\,000 \times 75 \times 250 = 7.9 \times 10^8$ N.
Boat weight = $mg = 8.0 \times 10^7 \times 9.81 = 7.8 \times 10^8$ N.
These are equal (within rounding errors).

23. Work done by a force

1 (a) $5000 \times 80 = 400\,000$ J

(b) Work is done on the brakes transferring kinetic energy from the car through heating.

2 $W = 30 \times 20 \times \cos 35° = 490$ J

24. Conservation of energy

(a) Chemical energy to kinetic energy (and gravitational potential energy if something is moved vertically, e.g. a car going uphill) and thermal energy.

(b) 25%

(c) 75% of the input energy goes to heat as the car is driving. The other 25% is mechanical work that the car has to do against the external frictional forces (air resistance, drag) as it moves through the air and along the road. This transfers the remaining 25% to thermal energy so that all of the chemical energy produced from the fuel is dissipated into the atmosphere through heating.

25. Kinetic and gravitational potential energy

(a) $h = 25$ m
$v = \sqrt{2gh} = \sqrt{2 \times 9.81 \times 25} = 22.1$ m s^{-1}

(b) The car will have to do work against friction as it moves from B to C. This means it will need more kinetic energy at B and a speed greater than 22.1 m s^{-1}.

(c) $h = 25 - 10 = 15$ m
$v = \sqrt{2gh}$
$= \sqrt{2 \times 9.81 \times 15} = 17.2$ m s^{-1}

26. Mechanical power and efficiency

$P = Fv$ so $F = \dfrac{P}{v} = \dfrac{50\,000}{2.0} = 25\,000$ N

28. Elastic and plastic deformation

(a) Up to 8.00 N the spring returns to its original length when the load is removed. However, when a weight of 12.00 N is suspended and removed the spring does not return to its original length; it undergoes a permanent plastic deformation.

(b) When the load is doubled from 4.00 N to 8.00 N the extension also doubles from 3.0 cm to 6.0 cm.

(c) $k = \dfrac{F}{x} = \dfrac{8.00}{0.060} = 133$ N m^{-1}

29. Stretching things

1 (a) A parallax error occurs if the experimenter's eye is above or below the level of the base of the load so that a value above or below the correct value is read.

(b) a systematic error

2 (a) The longer the wire, the larger the extension, so the effect of measurement errors is reduced and accuracy is improved.

(b) a metre rule

(c) a micrometer screw gauge

(d) Make several measurements, not just one. Make sure that parallax errors are avoided when reading the length. Measure the thickness of the wire in several places, not just one.

30. Force–extension graphs

(a) The graphs are all straight lines that pass through the origin.

(b) $k = \dfrac{F}{x} = \dfrac{8.0}{0.06} = 130$ N m^{-1}

(c) Work done = $(0.5 \times 24.0 \times 0.06) - (0.5 \times 20.0 \times 0.05)$
$= 0.22$ J

31. Stress and strain

1 (a) $\varepsilon = \dfrac{(1.81 - 1.80)}{1.80} = 5.6 \times 10^{-3}$

(b) $\sigma = \dfrac{F}{A} = \dfrac{6.2}{\left(\pi \times \left(\frac{0.0033}{2}\right)^2\right)} = 7.25 \times 10^5$ Pa

2 $F = \sigma_{uts}A = 2.20 \times 10^8 \times \pi \times \left(\dfrac{0.0002}{2}\right)^2 = 6.91$ N

32. Stress–strain graphs and the Young modulus

1 $E = \dfrac{\sigma}{\varepsilon} = \dfrac{550 \times 10^6}{0.0050} = 1.1 \times 10^{11}$ Pa = 110 GPa.

2 $E = \dfrac{\sigma}{\varepsilon} = \dfrac{Fl}{Ae}$, therefore $e = \dfrac{mgl}{AE} = \dfrac{mgl}{\pi\left(\dfrac{d}{2}\right)^2 E} = 0.017\,\text{m}$

33. Measuring the Young modulus

(a) Measure the gradient and calculate $E = \dfrac{l}{A} \times$ gradient. It gives a value of about 809 GPa.

(b) Measure the extension using a pointer attached to the wire and read directly above the scale to avoid parallax errors. Measure the diameter of the wire in several places using a micrometer screw gauge and take a mean.

35. Newton's laws of motion

Newton's third law: as the ball bearing moves downward it exerts a downward force on the fluid beneath it. The fluid exerts an equal upward force on the ball bearing. As it moves faster, this force increases.

Newton's second law: At first the weight of the ball bearing is greater than the reaction force (drag) acting on it from the fluid so there is a resultant force acting downward. The ball accelerates downward.

Newton's first law: the drag force increases with the ball bearing's speed. Eventually the drag is equal to the ball bearing's weight so the resultant force is zero. When the resultant force is zero there is no acceleration so the ball bearing continues to move at this terminal velocity.

36. Linear momentum

1 $p = 1200 \times v = 3000 \times 10$, so $v = 25\,\text{m s}^{-1}$

2 When a child falls onto the surface, the child's momentum changes rapidly to zero and the child it experiences a large force from the surface. A rubberised surface distorts during the collision, giving way under the child so that the stopping time ($\triangle t$) is increased. This reduces the rate of change of linear momentum and so reduces the force and therefore the risk of injury.

37. Impulse

(a) $5.0\,\text{kg m s}^{-1}$

(b) $F\triangle t = 5.0\,\text{N s}$ so $F = \dfrac{5.0}{2.0} = 2.5\,\text{N}$

(c) $p = mv = 5.0$ so $v = \dfrac{5.0}{0.100} = 50\,\text{m s}^{-1}$

(d) If fired vertically, there will be downward force of 1.0 N due to the rocket's mass. This would reduce the final velocity. Air resistance would also reduce the final velocity. In addition, the mass of the rocket would decrease as fuel is burnt. This would increase the final velocity.

38. Conservation of linear momentum – collisions in one dimension

Momentum is conserved so $60 \times 2.0 = 0.40 \times m$.
Therefore $m = 300\,\text{kg}$.

39. Collisions in two dimensions

(a) $1400 \times 25\cos(20°) = 32\,900\,\text{kg m s}^{-1}$

(b) $1400 \times 25\sin(20°) = 12\,000\,\text{kg m s}^{-1}$

40. Elastic and inelastic collisions

(a) Linear momentum of toy cannon $= m_C v_C = 0.2130 \times 0.35 = 0.075\,\text{kg m s}^{-1}$
Velocity of ball $= \dfrac{p_B}{m_B} = \dfrac{0.075}{0.0027} = 28\,\text{m s}^{-1}$

(b) Initial kinetic energy = 0.
Kinetic energy after firing $= \frac{1}{2}m_C v_C^2 + \frac{1}{2}m_B v_B^2$
$= 0.5 \times 0.2130 \times (0.35)^2 + 0.5 \times 0.0027 \times 28^2 = 1.0\,\text{J}$
This energy has been transferred from the spring of the toy into kinetic energy of the ball.

(c) The ball is so light that the impulse ($mv - mu = F\triangle t$) and thus the force that it imparts when it decelerates on impact is small. (In addition, the drag exerted by air resistance will decelerate it rapidly before it has travelled far.) (Other arguments, positive or negative, accepted as long as reasoning is sound.)

41. Electric charge and current

1 $I = \dfrac{\triangle Q}{\triangle t} = \dfrac{10 \times 10^{-6}}{10^{-3}} = 10\,\text{mA}$

2 $\triangle t = \dfrac{\triangle Q}{I} = \dfrac{1.25}{50 \times 10^3} = 25\,\mu\text{s}$

42. Charge flow in conductors

(a) Number of Cu^{2+} ions $= \dfrac{12}{(2 \times 1.60 \times 10^{-19})}$
$= 3.75 \times 10^{19}$ ions

(b) Number of ions $= \left(\dfrac{1.0}{63.5}\right) \times 6.02 \times 10^{23} = 9.5 \times 10^{21}$
Charge that must flow $= 9.5 \times 10^{21} \times 2 \times 1.60 \times 10^{-19}$
$= 3033\,\text{C}$
Time taken $\triangle t = \dfrac{\triangle Q}{I} = \dfrac{3033}{0.040} = 75\,800\,\text{s} = 21$ hours

43. Kirchhoff's first law

Known current entering junction $= 120 + 220\,\text{mA} = 340\,\text{mA}$
Known current leaving junction $= 60\,\text{mA}$
I_2 must therefore also be leaving junction (flow to right) to balance total current entering
$I_2 = 340 - 60 = 280\,\text{mA}$

44. Charge carriers and current

(a) $I = \dfrac{Q}{t}$, so in 1 s $Q = 0.250\,\text{C}$; $N = \dfrac{Q}{e} = \dfrac{0.250}{1.60 \times 10^{-19}}$
$= 1.6 \times 10^{18}$ electrons per second.

(b) $v = \dfrac{I}{Ane} = \dfrac{I}{\pi\left(\dfrac{d}{2}\right)^2 ne}$
$= \dfrac{0.250}{\left[\pi\left(\dfrac{0.25 \times 10^{-3}}{2}\right)^2 \times 8.5 \times 10^{28} \times 1.60 \times 10^{-19}\right]}$
$= 3.7 \times 10^{-4}\,\text{m s}^{-1}$

(c) 1.5 hours

(d) Electrons do not have to travel all the way from the cell in order to light the lamp. There are charge carriers in the wires and the lamp filament. As soon as the circuit is complete they all start moving, so the filament lights because the electrons already present inside it begin to move.

45. Electromotive force and potential difference

(a) 1.5 J

(b) $W = \varepsilon It = 1.5 \times 0.200 \times 5.0 \times 60 = 90\,\text{J}$

46. Resistance and Ohm's law

Component	Potential difference	Current	Resistance
W	40 V	0.20 A	200 Ω
X	6.0 V	50 mA	120 Ω
Y	600 V	30 mA	20 kΩ
Z	3.0 kV	60.0 μA	50 MΩ

47. *I–V* characteristics

If the ammeter is ideal then it will make no difference. A perfect ammeter has zero resistance, so there is no p.d. across it when current flows through it. This means that the voltmeter will read the same value as if it had been connected directly across the component. However, if the ammeter has some internal resistance then the voltmeter reading will be larger than for the component alone, so the value for the resistance will be incorrect.

48. Resistance and resistivity

1 $R = \dfrac{\rho l}{A} = \dfrac{150 \times 10^{-8} \times 0.20}{\left[\pi\left(\dfrac{0.50 \times 10^{-3}}{2}\right)^2\right]} = 1.53\,\Omega$

2 $R = \dfrac{\rho l}{A}$ so $A = \dfrac{\rho l}{R} = \dfrac{5.60 \times 10^{-8} \times 0.10}{4.0} = 1.4 \times 10^{-9}\,\mathrm{m}^2$

49. Resistivity and temperature

(a) Increases almost linearly.
(b) It is greater. It increases at a greater rate.

50. Electrical energy and power

1 (a) $I = \dfrac{P}{V} = \dfrac{2000}{230} = 8.7\,\mathrm{A}$

   (b) $E = 2 \times \dfrac{5}{60} = 0.167\,\mathrm{kWh}$, so cost $= 0.167 \times 20\mathrm{p} = 3.3\mathrm{p}$.

2 $\frac{1}{2}mv^2 = qV$ so $V = \dfrac{\frac{1}{2}mv^2}{q}$

   $= 0.5 \times 9.1 \times 10^{-31} \times \dfrac{(0.05 \times 3.0 \times 10^8)^2}{(1.6 \times 10^{-19})} = 640\,\mathrm{V}$

51. Kirchhoff's laws and circuit calculations

P.d. across one resistor in central branch $V = IR = 0.02 \times 200$ $= 4\,\mathrm{V}$, so p.d. across central branch is 8 V.
By Kirchhoff's second law this is equal to the e.m.f. so $\varepsilon = 8\,\mathrm{V}$.
By Kirchhoff's second law the p.d. across the lower resistor is also 8 V. The current through the lower resistor is

$I = \dfrac{V}{R} = \dfrac{8}{200} = 0.04\,\mathrm{A} = 40\,\mathrm{mA}$

By Kirchhoff's first law the current $I = 20\,\mathrm{mA} + 40\,\mathrm{mA} = 60\,\mathrm{mA}$

52. Resistors in series and parallel

1 Central series branch has resistance $150\,\Omega$. Three branches in parallel have resistance R_{para} where $\dfrac{1}{R_{\mathrm{para}}} = \dfrac{1}{150} + \dfrac{1}{50} + \dfrac{1}{100}$

$= \dfrac{11}{300}$ so $R_{\mathrm{para}} = \dfrac{300}{11} = 27.3\,\Omega$. Total circuit resistance is

therefore $80 + 27.3 = 107.3\,\Omega$. Current $I = \dfrac{V}{R} = \dfrac{12}{107.3} = 0.11\,\mathrm{A}$
$= 110\,\mathrm{mA}$.

53. DC circuit analysis

1 Total circuit resistance $= 3.33\,\mathrm{k}\Omega$
Current from cell $= \dfrac{12}{3333} = 0.0036\,\mathrm{A}$
P.d. across top $2.0\,\mathrm{k}\Omega$ resistor $= 0.0036 \times 2000 = 7.20\,\mathrm{V}$
P.d. across both parallel arms $= 12.0 - 7.20 = 4.80\,\mathrm{V}$
P.d. across $3.0\,\mathrm{k}\Omega$ resistor $= \frac{3}{4} \times 4.80 = 3.6\,\mathrm{V}$

2 The central resistor has no p.d. across it so no current flows in it and it can be ignored when analysing this circuit. Each of the remaining parallel branches has a total resistance of $10\,\Omega$ so draws $\frac{20}{10} = 2.0\,\mathrm{A}$. The total current is therefore 4.0 A. (The total circuit resistance is $5.0\,\Omega$.)

54. E.m.f. and internal resistance

(a) $\varepsilon = 12\,\mathrm{V}$, $r = 0.50\,\Omega$ (same method as in worked example)
(b) $V = IR = \dfrac{\varepsilon R}{(R + r)} = \dfrac{12.0 \times 2.5}{(3.0)} = 10.0\,\mathrm{V}$
(c) Maximum possible current would be drawn with a circuit resistance of 0 (a short circuit)
$\varepsilon = IR + Ir$ so $\varepsilon = I_{\max}r$; $I_{\max} = \dfrac{12.0}{0.50} = 24\,\mathrm{A}$

55. Experimental determination of internal resistance

$\varepsilon = 3.0\,\mathrm{V}$, $r = 1.5\,\Omega$

56. Potential dividers

(a) $V_{\mathrm{out}} = 20 \times \left(\dfrac{250}{300}\right) = 16.7\,\mathrm{V}$
(b) $V_{\mathrm{out}} = 20 \times \left(\dfrac{125}{175}\right) = 14.3\,\mathrm{V}$

57. Investigating potential divider circuits

(a) 3.3 V in light, 5.0 V in dark
(b) 0.45 V in light, 4.2 V in dark
(c) 0.025 V in light, 1.0 V in dark

59. Properties of progressive waves

Transverse: any of the EM spectrum, seismic S-waves (accept water waves, although these are actually more complex)
Longitudinal: sound, ultrasound, seismic P-waves, compression waves on a spring or slinky

60. The wave equation

1 $f = c\lambda = \dfrac{3.00 \times 10^8}{10^{-10}} = 3.0 \times 10^{18}\,\mathrm{Hz}$.

2 Frequency is unchanged. The number of waves per second entering the material is the same as the number of waves per second inside the material (they just bunch up). The wavelength reduces by the same factor as the velocity since $v = f\lambda$ and f is constant.

3 If distance in km is x then time for each type of wave is:
$t_P = \dfrac{x}{6}$ and $t_S = \dfrac{x}{4.5}$.
$t_S - t_P = 15$. Substituting and solving gives $x = 270\,\mathrm{km}$.

61. Graphical representation of waves

(a) $f = \dfrac{1}{T} = \dfrac{1}{5.0} = 0.20\,\mathrm{Hz}$
(b) $\lambda = 40\,\mathrm{m}$. The swimmers are in antiphase ($180°$ out of phase) on the same wave so they are separated by half a wavelength.
(c) $v = f\lambda = 0.20 \times 40 = 8.0\,\mathrm{m\,s^{-1}}$

62. Using an oscilloscope to display sound waves

(a) $T = 5.0 \times 100\,\mu\mathrm{s} = 500\,\mu\mathrm{s}$
(b) $f = \dfrac{1}{T} = \dfrac{1}{500 \times 10^{-6}} = 2000\,\mathrm{Hz}$
(c) $A = 3.2 \times 50\,\mathrm{mV} = 160\,\mathrm{mV}$

63. Reflection, refraction and diffraction

1 B

2 The wavelengths of audible sounds are comparable to or longer than the diameter of the speaker, so a great deal of diffraction occurs, spreading the sounds to large angles. The wavelength of visible light is tiny compared with the diameter of the torch so there is hardly any diffraction and the light rays travel straight ahead, forming a narrow beam.

64. The electromagnetic spectrum

1 $f = \frac{c}{\lambda} = 4.3 \times 10^{14}\,\text{Hz}$

2 The area surrounding the point source at distance r is the surface of a sphere of radius r: $4\pi r^2$. The intensity at that distance is $\frac{P}{A} = \frac{P}{4\pi r^2}$ so $P \propto \frac{1}{r^2}$.

65. Polarisation

Light transmitted by the first filter falls onto the middle filter at a polarisation angle of 45°, so this filter transmits a reduced intensity, polarised in this direction. This light falls on the last filter at 45° to its polarising direction so again the intensity is reduced but some light is transmitted, and this will be polarised parallel to the polarising direction of the final filter.

66. Refraction and total internal reflection of light

1 $\frac{\sin 43°}{\sin \theta_2} = 1.50$; $\sin \theta_2 = \frac{\sin 43°}{1.50} = 0.455$.
Angle of refraction = 27°.

2 The critical angle for water, C: $\sin C = \frac{1}{1.33} = 0.752$,
so $C = 49°$. The incident angle of 52° is greater than the critical angle so the ray undergoes total internal reflection and there is no refracted ray.

3 $n = \frac{1}{\sin C} = \frac{1}{\sin 25°} = 2.4$

67. The principle of superposition

(a) $x = 0$, $y = +0.8$; $x = 1$, $y = 1.8$; $x = 2$, $y = 1.3$; $x = 3$, $y = -0.3$; $x = 4$, $y = -1.75$ (approx)

(b) 1.7−1.9 m

(c) Diagram should be like the one below:

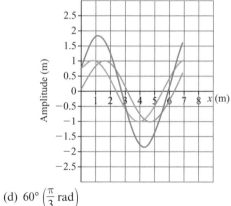

(d) 60° $\left(\frac{\pi}{3}\,\text{rad}\right)$

68. Interference

1 Sound has to travel further from one of the speakers than from the other. The amplitude of sound falls with distance from the speaker because it is spreading out into a larger space. This means that whilst the sounds superpose in antiphase they do not cancel completely; one is louder than the other.

2 Light from car headlamps is not monochromatic; it contains the full visible spectrum, a wide range of wavelengths. If interference did occur the spacings from each wavelength would be different so the maxima and minima would overlap everywhere and we would not be able to see a clear interference pattern. However, the light from two headlamp bulbs is incoherent so no consistent interference pattern can be formed even for a single wavelength. This results in the intensities adding and a general patch of illumination with no interference pattern.

69. Two-source interference and the nature of light

1 $\lambda = \frac{ax}{D}$ and $x = \frac{1.19\,\text{cm}}{4}$, $\lambda = \frac{1.19 \times 10^{-2} \times 0.42 \times 10^{-3}}{(4 \times 2.40)}$
= $5.21 \times 10^{-7}\,\text{m} = 521\,\text{nm}$

2 The double slit experiment produced an interference pattern. Interference occurs because of the principle of superposition. This is a property of waves.

70. Experimental determination of the wavelength of light

1 $\lambda = \frac{(0.22 \times 10^{-3} \times 5.00 \times 10^{-2})}{(15 \times 1.25)} = 5.87 \times 10^{-7}\,\text{m}$.
Fractional uncertainty $= \frac{0.01}{1.25} + \frac{0.02}{0.22} + \frac{0.05}{5.00} = 0.11\ (11\%)$
$\lambda = 587 \pm 64\,\text{nm}$

2 (a) Monochromatic light is of one wavelength. If more than one wavelength is present then there will be maxima at different spacings for each wavelength and the pattern will not be clear.
   (b) If the light is incoherent there will be no consistent interference and no fringes will form.

71. Stationary waves

(a) Distance between nodes is 0.40 m so wavelength is 0.80 m.
(b) $v = f\lambda = 220 \times 0.80 = 176\,\text{m s}^{-1}$
(c) Same amplitude but 180° phase difference.
(d) The waves superpose in antiphase at this point so there is destructive interference. Both waves have the same amplitude so they cancel, resulting in zero amplitude.
(e) Progressive waves from the vibrator travel along the wire and reflect at the end. Superposition of the waves travelling to the right and to the left results in the formation of nodes and antinodes in fixed positions − a stationary wave.

72. Stationary waves on a string

(a) $v = f\lambda = 110 \times 2 \times 0.65 = 143\,\text{m s}^{-1}$
(b) $4f_1 = 4 \times 110 = 440\,\text{Hz}$
(c) The strings will expand slightly so the tension will drop. This reduces the speed of transverse waves in the string. Since $f = \frac{v}{\lambda}$ and the wavelengths of the harmonics are unchanged, all the harmonic frequencies are reduced. The guitar sounds flat.
(d) Frequency of the second harmonic is 220 Hz. In air
$\lambda = \frac{v}{f} = \frac{340}{220} = 1.55\,\text{m}$.

73. Stationary sound waves

(a) At displacement antinodes the air particles are vibrating with their largest amplitude so the sawdust is pushed away from these positions and accumulates at the nodes.
(b) The distance between adjacent antinodes is $\frac{\lambda}{2}$ so
$3 \times \frac{\lambda}{2} = 42\,\text{cm}$ and $\lambda = 28\,\text{cm}$.
(c) $f = \frac{v}{\lambda} = \frac{340}{0.28} = 1200\,\text{Hz}$.

74. Stationary waves in closed and open tubes

(a) (i) $f = \dfrac{v}{\lambda} = \dfrac{v}{2L} = \dfrac{340}{1.80} = 189\,\text{Hz}$

(ii) $f = \dfrac{v}{4L} = 94\,\text{Hz}$

(iii) $f = \dfrac{v}{2L} = 189\,\text{Hz}$

(b) Fundamental frequency = 94 Hz. Next two are the third and fifth harmonics: 283 Hz, 472 Hz.

76. The photoelectric effect

1 Any two superposition effects, for example the formation of interference fringes in a double slit experiment or when using a diffraction grating; the diffraction pattern formed when light passes through a narrow slit.

2 Intensity determines how much energy is delivered per unit area to the surface. If it takes a certain amount of energy to eject an electron, then higher intensities ought to do this more effectively. In the wave theory there is no direct link between wave energy and frequency.

3 There is a threshold frequency for each metal, and UV light has a higher frequency than red light. The frequency of red light must be below this threshold.

77. Einstein's photoelectric equation

(a) $\phi = 2.5 \times 1.60 \times 10^{-19} = 4.0 \times 10^{-19}\,\text{J}$

(b) $\phi = hf_0$ so $f_0 = \dfrac{\phi}{h} = \dfrac{4.0 \times 10^{-19}}{6.63 \times 10^{-34}} = 6.03 \times 10^{14}\,\text{Hz}$

(c) $KE_{max} = hf - \phi = \dfrac{hc}{\lambda} - \phi = \dfrac{6.63 \times 10^{-34} \times 3.00 \times 10^8}{(420 \times 10^{-9})}$

$- 4.0 \times 10^{-19} = 0.04736 \times 10^{-17} - 4.0 \times 10^{-19} = 7.36 \times 10^{-20}\,\text{J}$

78. Determining the Planck constant

Plot V (y-axis) against $\dfrac{1}{\lambda}$ (x-axis) $h = \text{gradient} \times \left(\dfrac{e}{c}\right)$

$= 6.63 \times 10^{-34}\,\text{J s}$ Allow values of $6.55 - 6.65 \times 10^{-34}\,\text{J s}$

79. Electron diffraction

(a) $\lambda = \dfrac{h}{\sqrt{(2meV)}} = 1.8 \times 10^{-10}\,\text{m}$

(b) $2.0\,\text{eV} = 3.2 \times 10^{-19}\,\text{J}$.

$KE = \frac{1}{2}mv^2 = \dfrac{(mv)^2}{2m}$, so $mv = \sqrt{(2m \times KE)}$ and

$\lambda = \dfrac{h}{\sqrt{(2m \times KE)}} = 8.7 \times 10^{-10}\,\text{m}$

(c) Both have a single charge so both gain the same amount of

energy (eV). However, $eV = \dfrac{p^2}{2m}$, where $p = mv$ (momentum).

Rearranging this gives $p = \sqrt{(2meV)}$, so the proton will have greater momentum and therefore shorter wavelength, from

$\lambda = \dfrac{h}{p}$.

80. Wave-particle duality

Even if we assume that the de Broglie relation applies to all matter, for macroscopic objects − for example the tennis ball − the wavelength is minute, and therefore the effects of diffraction etc., are vanishingly small.

The de Broglie equation applied to a tennis ball of mass 200 g travelling at 50 m s^{-1} would give a value for the wavelength:

$\lambda = \dfrac{h}{mv} = \dfrac{6.63^{-34}}{0.2 \times 50} = 6.63 \times 10^{-35}\,\text{m}$

This is infinitesimal compared tot he size of the ball, so the wave effects are virtually non-existent and can be ignored.

A level

82. Temperature and thermal equilibrium

Introduce smoke into a small glass container, which is illuminated using a bright lamp. The container is then observed through a microscope. The observer will see a large number of points of light which appear to be moving in a series of random jumps. The evidence this experiment provides for the movement of molecules in a fluid is:

- The points of light are smoke particles. They are moving in a random jerky fashion because they are being buffeted by particles of air.
- The particles of smoke must be large in comparison with the wavelength of light in order for them to be illuminated by the light. The air particles, however, are much smaller.
- Continuous (though jerky) movement of the smoke particles indicates continuous movement of the particles of air.
- Because the smoke particles are so much bigger, motion must be caused by many air molecules hitting each smoke particle.

83. Solids, liquids and gases

Energy is transferred to the ice, which increases the intermolecular potential energy. As the intermolecular potential energy increases the molecules move farther apart and the ice melts. Kinetic energy remains constant during melting, hence the temperature stays constant.

84. Specific heat capacity

$c = \dfrac{E}{m\triangle\theta}$ so $E = cm\triangle\theta$

with $c = 4180$, $m = 0.245$ and $\triangle\theta = 95 - 12.5 = 82.5$

Assuming that the water is heated evenly and that the kettle is insulated against heat loss, the energy required is:

$E = 4180 \times 0.245 \times 82.5 = 8.45 \times 10^4\,\text{J}$

85. Specific latent heat 1

(a) $L = \dfrac{E}{m} = \dfrac{(1000 \times 4.3 \times 60)}{0.8} = 322\,500\,\text{J} = 323\,\text{kJ kg}^{-1}$

(b) This experiment uses crushed ice to ensure that as much of the ice as possible is in contact with the heater.

(c) If the mass of ice melted by energy transfer from the surroundings is not accounted for in the calculation, then the mass of ice melted by the heater will be overestimated. Hence the value of the specific latent heat calculated will be too small, since latent heat is inversely proportional to mass.

(d) Possible answers:
- Measuring errors
- No control considered
- Ice not completely crushed
- Rounding errors

(Students to choose two)

86. Specific latent heat 2

When water reaches the vaporisation temperature of 100°C, it remains at that temperature even though heat energy is still being added. This energy is used to separate the water molecules to produce steam − the *latent heat energy*. Thus, although the temperature is the same in boiling water and steam, the steam actually possesses more heat energy than the water. When the steam condenses on the skin this excess heat energy is transferred to the skin, so a scald with steam will be worse than with water.

87. Kinetic theory of gases

Pressure difference $P_2 - P_1 = \dfrac{F_2}{A} - \dfrac{F_1}{A} = \dfrac{F_2 - F_1}{A}$

Therefore,

Net force = pressure difference × area = $1500 \times 2.5 = 3750\,\text{N}$

88. The gas laws: Boyle's law

When it is inflated, the air pressure *inside* the balloon must be equal to the atmospheric pressure *outside*, otherwise it would carry on either inflating or deflating until this were the case.

$pV = k$ (constant) So $k = 102 \times 2 = 204$

Thus, when $p = 95$ kPa, $V = \frac{204}{95} = 2.147$ m^3.

As the balloon rises, the atmospheric pressure will decrease, so the pressure in the fixed mass of air inside the balloon must also decrease. According to Boyle's law, pressure is inversely proportional to volume, so as the balloon continues to rise, the decrease in atmospheric pressure will be accompanied by an increase in the volume of the balloon. The volume of the balloon increases until the balloon skin has been stretched to its limit; at this point the balloon bursts. (We have assumed that temperature remains constant, although in reality this will not be the case.)

89. The gas laws: the pressure law

1   As the temperature increases, the kinetic energy of the air molecules increase. Hence the molecules move at higher speed with more momentum. Therefore there are more collisions between the molecules and the container walls per unit time, and a greater rate of change of momentum per collision. Hence the force exerted over a given area of the walls is increased and so the pressure increases.

2   According to Gay–Lussac's law, for a given volume of air:

$$\frac{p_1}{T_1} = \frac{p_2}{T_2}$$

Where p_n and T_n are the pressure and temperature (in kelvins) at a particular time

Remembering that 0 K $= -273$ °C, we have:

$$p_2 = \frac{116 \times 313}{273} = 133 \text{ kPa}$$

90. The equation of state of an ideal gas

1   From observation, there is a clear linear relationship between temperature and volume. Assuming that this is a universal relation, the line can be extended back to the point where the volume occupied by the gas is zero. This should be at about $-273°$ on the temperature axis.

2   Assuming that nitrogen behaves as an ideal gas and using the equation of state:

1000 cm$^3 = 0.001$ m^3

$$n = \frac{pV}{RT} = \frac{1.01 \times 10^5 \times 0.001}{8.31 \times 273} = 0.0445 \text{ moles}$$

91. The kinetic theory equation

1   Reducing the volume occupied by the gas whilst maintaining a constant temperature increases the collision rate of the molecules with the container walls. The greater the collision rate of molecules with the container walls the greater the rate of change of momentum and the greater the force exerted on a given area, hence the greater the pressure.

2   (a) Assuming that the Helium behaves as an ideal gas, we first calculate how many moles (n) we have. Then using the equation of state and the information given in the example calculate the pressure:

$$\text{moles} = \frac{\text{molar mass}}{\text{mass}} = \frac{4}{2.5} = 0.625 \text{ mol}$$

Now applying the equation of state for the temperature T = 303K

(b) $pV = \frac{1}{3}Nm\overline{c^2}$

$$c_{rms} = \sqrt{\overline{c^2}} = \sqrt{\frac{3pV}{Nm}} = \sqrt{\frac{3 \times 109 \times 10^3 \times 0.0145}{2.5 \times 10^{-3}}} = 1380 \text{ ms}^{-1}$$

92. The internal energy of a gas

(a) As the balloon heats up, the heat energy is converted to kinetic energy in the gas molecules. Using the expression for molecular kinetic energy, an increase in E_k, with no change in m, means that the value for $\overline{c^2}$ must increase accordingly. So the mean square velocity of the gas particles will increase with heat, meaning that more particles will be hitting the walls of the balloon in a given time and the pressure exerted by the gas will increase.

(b) $T \propto \frac{1}{2}m\overline{c^2}$ $\therefore \dfrac{\overline{c_2^2}}{\overline{c_1^2}} = \dfrac{T_2}{T_1}$

$$\therefore \frac{\overline{c_2^2}}{\overline{c_1^2}} = \frac{103}{100} = \frac{\theta + 273}{18 + 273}$$

$103 \times 291 = \theta + 273$

$\theta = 27$ °C

94. Angular velocity

(a) $\omega = \dfrac{\triangle\theta}{\triangle t} = \dfrac{2\pi}{5} = 0.4\pi$ rad s^{-1}

(b) Comparing the speed (tangential velocity) of the inner and outer children:

$v_1 = \omega r_1 = 1.2 \times 0.4\pi = 0.48\pi$ m s^{-1}

$v_2 = \omega r_2 = 2.5 \times 0.4\pi = \pi$ m s^{-1}

$\dfrac{v_1}{v_2} = \dfrac{0.48\pi}{\pi} = 0.48$

95. Centripetal force and acceleration

$\omega = \dfrac{2\pi}{T} = \dfrac{2\pi}{30 \times 60} = 3.5 \times 10^{-3}$ rad s^{-1}

$a = \omega^2 r = (3.5 \times 10^{-3})^2 \times 60 = 7.3 \times 10^{-4}$ m s^{-2}

96. Simple harmonic motion

(a) The single condition which must be met for an object to be said to move with simple harmonic motion is as follows: When the body is displaced from equilibrium, there must be a restoring force, F, directed towards the equilibrium point, which produces an acceleration directly proportional to the displacement, x, of the body from the equilibrium point.

(b) Free body diagram drawn for pendulum displaced through a small angle. Weight of bob resolved into force along and perpendicular to tension in string. Statement that component of weight perpendicular to tension only approximately acts towards the equilibrium position.

97. Solving the s.h.m. equation

$$T = 2\pi\sqrt{\frac{m}{k}} = 2\pi\sqrt{\frac{1280}{30\,000}} = 1.3 \text{ s}$$

The period of oscillation is 0.96 seconds

98. Graphical treatment of s.h.m.

$a = -\omega^2 x$ $\therefore 2.2 \times 10^{-2} = \omega^2 \times 4.5 \times 10^{-2}$

$$\omega = \sqrt{\frac{2.2}{4.5}} = 0.7 \text{ rad s}^{-1}$$

$$v_{max} = \omega A = 0.7 \times 4.5 \times 10^{-2} = 3.2 \times 10^{-2} \text{ m s}^{-1}$$

99. Energy in s.h.m.

$x = A\cos\omega t$ so when $x = \frac{1}{2}A$, $\omega t = \frac{\pi}{3}$

Kinetic energy: $E_k = \frac{1}{2}m\omega^2 A^2 \sin^2\omega t$

Potential energy: $E_p = \frac{1}{2}m\omega^2 A^2 \cos^2\omega t$

Thus $\dfrac{E_k}{E_p} = \dfrac{\sin^2\left(\frac{\pi}{3}\right)}{\cos^2\left(\frac{\pi}{3}\right)} = 3$

100. Forced oscillations and resonance

As the car travels over the uneven surface of a road, the passengers do not want to experience large displacement, but they also don't want to be able to feel every bump. The suspension system should be fairly heavily damped, but with some give.

102. Gravitational fields

1 Newtons (N) are the S.I. unit of force. From Newton's first law, $F = ma$, we know that we can rewrite the unit as:
$$N = kg\,m\,s^{-2}$$
Therefore
$$N\,kg^{-1} = \frac{kg\,m\,s^{-2}}{kg} = m\,s^{-2}$$

2 From worked example, $g_{Mars} = 3.83\,N\,kg^{-1}$
Taking $g_{Earth} = 9.81\,N\,kg^{-1}$
Astronaut mass $= 730 \div 9.81 = 74.4\,kg$
Pack mass $= 35\,kg$
Thus astronaut weight on Mars $= 74.4 \times 3.83 = 285\,N$
Pack weight on Mars $= 35 \times 3.83 = 134\,N$

103. Newton's law of gravitation

$$g = -\frac{GM}{r^2}$$
$$9.81 = -\frac{GM}{(6\,400\,000)^2}$$
$$8.81 = -\frac{GM}{r^2}$$
$$\frac{9.81}{8.81} = \frac{r^2}{(6\,400\,000)^2}$$
$$r = \sqrt{(6\,400\,000)^2 \times \frac{9.81}{8.81}} \quad r = 6\,750\,000\,m$$

At Earth's surface $r = 6400\,km$ so to calculate the distance one would have to travel upwards (h) to reduce the gravitational field by $1\,Nkg^{-1}$ we must take the difference between the Earth's radius and the new radius we have calculated:
$$h = 6\,750\,000 - 6\,400\,000 = 350\,00\,m = 350\,km$$

We can see that the strength of Earth's gravitational field is inversely proportional to the square of it's radius. We have just calculated that to reduce the field strength by approximately 10% we would have to travel 350 km above Earths surface as such close to the surface (say within 30 km) the gravitational field strength would not change by more than 1%.

104. Kepler's laws for planetary orbits

Kepler's 3rd law: $T^2 = \left(\dfrac{4\pi^2}{GM}\right)r^3$

We know that $r_{Mars} = 1.52 \times r_{Earth}$ while G, M are constant.
Also that $T_{Earth} = 365.25$ days
So $T^2_{Mars} = 3.51 \times T^2_{Earth} = 3.51 \times 133\,408 = 468\,502$
Thus $T_{Mars} = 684$ days
Comment: approximate value due to average distances

105. Satellite orbits

For Mars, $T = 24$ hours and 37 minutes $= 88620\,s$

So, using $T^2 = \left(\dfrac{4\pi^2}{GM}\right)r^3$,

$$r = \sqrt[3]{\frac{6.67 \times 10^{-11} \times 6.4 \times 10^{23} \times 88\,620^2}{4\pi^2}}$$

Giving a value for the orbital radius $r = 20\,400\,km$
This is 17 000 km above the surface of Mars.

106. Gravitational potential

$$V_1 = \frac{-(6.67 \times 10^{-11} \times 5.98 \times 10^{24})}{6.36 \times 10^6} = -\frac{39.89}{6.36} \times 10^7 = -6.27 \times 10^7$$
$$V_2 = \frac{-(6.67 \times 10^{-11} \times 5.98 \times 10^{24})}{4.22 \times 10^7} = -\frac{39.89}{4.22} \times 10^6 = -9.46 \times 10^6$$
$\therefore$ change in gravitational potential $= 6.27 \times 10^7 - 9.46 \times 10^6$
$$= 5.33 \times 10^7\,J\,kg^{-1}$$

107. Gravitational potential energy and escape velocity

1 $\triangle E_{gp} = GMm\left(\dfrac{1}{r_1} - \dfrac{1}{r_2}\right) = 6.67 \times 10^{-11} \times 5.98 \times 10^{24} \times m$
$$\times \left(\frac{1}{6399 \times 10^3} - \frac{1}{6360 \times 10^3}\right)$$
$$= -3.8 \times 10^5 \times m\,J$$
$\triangle E_{gp} = mg\triangle h = -m \times 9.81 \times 39 \times 10^3 = -3.8 \times 10^5 \times m\,J$
The approximation still works well for this change in height.

2 $\triangle E_{gp} = GMm\left(\dfrac{1}{r_1} - \dfrac{1}{r_2}\right)$
$$= 6.7 \times 10^{-11} \times 5.98 \times 10^{24} \times 5950$$
$$\times \left(\frac{1}{6.655 \times 10^6} - \frac{1}{4.216 \times 10^7}\right)$$
$$= 3.02 \times 10^{11}\,J$$

109. Formation of stars

1 Jupiter is more massive than the Earth but it is also much larger – the diameter of Jupiter is around 11 times that of Earth. Standing on the 'surface' of Jupiter you would be about 10 times farther from the centre of the planet than you would be if you were standing on Earth.
The force due to gravity is directly proportional to the mass of the planet, but is in inverse proportion to the square of the distance from the planetary centre. Thus it is $\frac{300}{100} = 3$ times as strong as on the surface of the Earth.
(The 'surface' of Jupiter – a gas giant – is generally taken to be the point in the planetary atmosphere where the pressure is approximately 1 atmosphere.)

2 $\triangle E_{gp} = GM_E m_M\left(\dfrac{1}{r_2} - \dfrac{1}{r_1}\right)$
$$= 6.67 \times 10^{-11} \times 5.97 \times 10^{24} \times 7.35 \times 10^{22}$$
$$\times \left(\frac{1}{3.84 \times 10^8} - \frac{1}{3.84 \times 10^8 + 10}\right)$$
$$= 2.93 \times 10^{37} \times 6.78 \times 10^{-17}$$
$$= 2.0 \times 10^{21}\,J$$

110. Evolution of stars

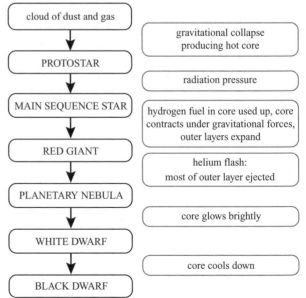

cloud of dust and gas

↓ gravitational collapse producing hot core

PROTOSTAR

↓ radiation pressure

MAIN SEQUENCE STAR

↓ hydrogen fuel in core used up, core contracts under gravitational forces, outer layers expand

RED GIANT

↓ helium flash: most of outer layer ejected

PLANETARY NEBULA

↓ core glows brightly

WHITE DWARF

↓ core cools down

BLACK DWARF

111. End points of stars

Planetary nebula – an expanding glowing shell of gas ejected from a red giant towards the end of its life

Supernova remnant – large glowing shell of stellar material ejected by an exploding star

112. The Hertzsprung–Russell diagram

Planetary nebulae and supernova remnants are often similarly shaped and sized however they are formed in different ways. Planetary nebulae are produced by red giants at the end of their life and consist of an expanding shell of glowing gas. A Supernovae remnant is the shell of stellar material produced at the end of a red supergiant's life after it has exploded.

113. Energy levels in atoms

Rutherford model: the electrons in an atom circle the nucleus at certain distances, like the planets around the sun. This model had no explanation for the formation of atomic spectra (the lines). Bohr's model: the electrons travelled around the nucleus in orbits with specific energy levels, and that they could jump from one level to another – but not anywhere in between. When they jump between levels, the electrons either emit or absorb packets of energy as photons of particular frequencies, creating atomic spectra.

114. Emission and absorption spectra

- Place a piece of opaque paper with a vertical slit in front of the discharge tube to form a narrow beam
- Place the diffraction grating between the slit and a sheet of photographic paper
- Switch off the lab lights
- Connect the discharge tube and switch on
- When the paper has been fully exposed to the emission spectrum, cover and develop the photograph

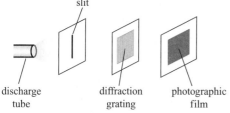

slit

discharge tube · diffraction grating · photographic film

- Repeat the experiment with the discharge tube replaced by a source of white light as a control.
- Compare the two spectra to evaluate where the absorption lines lie.
- Using a standard reference spectrum the wavelengths in the emission spectrum can be evaluated.

115. Wien's law and Stefan's law

1 Stefan's law: $L = A\sigma T^4$, so $A = \dfrac{L}{\sigma T^4}$

With $L = 3.6 \times 10^{26}$ W, $T = 5800$ K
and $\sigma = 5.67 \times 10^{-8}$ W m^{-2} K^{-4}, we have
$$A = \frac{3.6 \times 10^{26}}{5.67 \times 10^{-8} \times 5800^4} = 5.61 \times 10^{18}$$
Thus the surface area of the sun is approximately 5.61×10^{18} m^2

2 $T = \dfrac{2.9 \times 10^{-3}}{\lambda_{max}} = \dfrac{2.9 \times 10^{-3}}{269 \times 10^{-9}} = 10\,780$ or $10\,800$ K

$L = 4\pi r^2 \sigma T^4$

so $r^2 = \dfrac{L}{4\pi\sigma T^4} = \dfrac{3.89 \times 10^{28}}{(4\pi \times 5.67 \times 10^{-8} \times (10\,780)^4} = 4.04 \times 10^{18}$

$r = 2.0 \times 10^9$ m

116. The distances to stars

Annual parallax is only useful for determining the distances to nearby stars because the farther away a star is, the smaller the parallax angle becomes, so the margin of error will overwhelm the measurement.

117. The Doppler effect

$$\frac{\triangle\lambda}{\lambda} \approx \frac{\triangle f}{f} \approx \frac{v}{c} \text{ so } v \approx \frac{c\triangle\lambda}{\lambda} = \frac{3.00 \times 10^8 \times 70 \times 10^{-9}}{635 \times 10^{-9}}$$
$$= 3.3 \times 10^7\,\text{m s}^{-1} \text{ (about } 1.2 \times 10^8\,\text{km h}^{-1})$$

118. Hubble's law

The main source of inaccuracy in calculating the Hubble constant is in the calculation of distance to distant galaxies using their red shift values. They are not all moving directly away from us and the angle at which we see their path will affect the value we can ascertain for z.

119. The evolution of the Universe

The formula $t = \dfrac{1}{H_0}$ for the age of the Universe is an approximation, as it assumes that Hubble's law is valid for all time since the Big Bang, and is derived assuming a galaxy has been moving with constant velocity outwards since $t = 0$. Also, inaccuracy in the current value of the Hubble constant will introduce inaccuaracy into the calculation.

120. Capacitors

(a) $W = \frac{1}{2}QV = \frac{1}{2}CV^2$
$125 = \frac{1}{2} \times (1200 \times 10^{-6}) \times V^2$
$V^2 = 208\,000$
$V \approx 500$ V
(b) $V = IR$
$500 = I \times 150$
$I = 3.3$ A

121. Series and parallel capacitor combinations

Total capacitance C is given by $\dfrac{1}{C} = \dfrac{1}{560} + \dfrac{1}{470} = \dfrac{1}{256}$
So $C = 256 \times 10^{-6}$ F
Energy: $W = \frac{1}{2}CV^2 = \frac{1}{2} \times 256 \times 10^{-6} \times 144 = 0.018$ J

122. Capacitor circuits

$\tau = CR = 4700 \times 10^{-6} \times 2.2 \times 10^3 = 10.34$ s
Fully charged in $\approx 5\tau = 52$ s

123. Exponential processes

$V = V_0 e^{-\frac{t}{RC}}$
$\dfrac{V}{V_0} = e^{-\frac{t}{RC}} = e^{-\frac{0.01}{2200}C}$
$\ln 0.85 = -\dfrac{0.01}{2200}C$
$C = -\dfrac{0.01}{2200} \times \dfrac{1}{0.162} = 2.8 \times 10^{-5}$ F $= 28\,\mu$F

125. Electric fields

[students' own diagrams similar to]

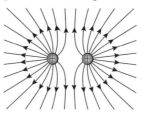

126. Coulomb's law

From $E = \dfrac{Q}{4\pi\varepsilon_0 r^2}$, $Q = E \times 4\pi\varepsilon_0 r^2$

Charge $= 350 \times 4\pi \times 8.85 \times 10^{-12} \times \left(\dfrac{0.45}{2}\right)^2 = 2.0 \times 10^{-9}\,\text{C}$

127. Similarities between electric and gravitational fields

Physicists model the forces between charges and masses by using the idea of a field. A gravitational field surrounds all masses, and an electric field surrounds all charges. The interaction of two gravitational fields results in an attractive force. The interaction of two electric fields results in a force which may be repulsive (between two like charges) or attractive (between two unlike charges). Displacing masses in a gravitational field results in an energy transfer. Physicists use this energy transfer to define the concept of gravitational potential. Similarly for charges in an electric field, electric potential can be defined.

128. Uniform electric fields

$C = \dfrac{\varepsilon_0 A}{d}$

with $\varepsilon_0 = 8.85 \times 10^{-12}$, $A = 16\,\text{cm}^2 = 1.6 \times 10^{-3}\,\text{m}^2$, $d = 5 \times 10^{-6}\,\text{m}^2$

So $C = \dfrac{1.416 \times 10^{-14}}{5 \times 10^{-6}} = 2.832 \times 10^{-9}\,\text{F}$

Capacitance $= 2.83\,\text{nF}$

$Q = CV$

With $C = 2.832 \times 10^{-9}$, $V = 9$

So $Q = 2.54 \times 10^{-8}$ coulombs

129. Charged particles in uniform electric fields

Remembering that the mass and charge for an electron are 9.11×10^{-31} and 1.6×10^{-19}, respectively, and the electrical field strength, E, is given as $3600\,\text{Vm}^{-1}$

So the acceleration is $a = \dfrac{qE}{m} = \dfrac{1.6 \times 10^{-19} \times 3600}{9.11 \times 10^{-31}}$

$\qquad\qquad\qquad\qquad = 6.32 \times 10^{14}\,\text{ms}^{-2}$

And the time spent in the field is given by

$t = \dfrac{0.05}{1.2 \times 10^7} = 4.17 \times 10^{-9}$ seconds

The deflection is then given by

$s = ut + \tfrac{1}{2}at^2 = 0 + \tfrac{1}{2} \times (6.32 \times 10^{14}) \times (4.17 \times 10^{-9})^2$
$\quad = 5.49 \times 10^{-3}\,\text{m}$

So when the electron leaves the region between the plates it is deflected by 5.49 mm from the horizontal.

130. Electric potential and electric potential energy

(a) When held at rest, the potential energy, E, of each droplet is given by:

$E = \dfrac{qQ}{4\pi\varepsilon_0 r} = \dfrac{1 \times 10^{-6} \times 1 \times 10^{-6}}{4 \times 3.14 \times 8.85 \times 10^{-12} \times 0.02} = 0.45\,\text{J}$

So the net kinetic energy of the system when the droplets are distant is 0.9 J

(b) When the droplets are at rest, their resultant momentum is $0\,\text{kg m s}^{-1}$.

From the law of conservation of momentum, the resultant momentum when they are far apart is $m_1 v_1 - m_2 v_2 = 0$ (with the minus sign necessary because the velocities are in different directions).

Thus $m_1 v_1 = m_2 v_2 =$ some constant k.

If the kinetic energy is shared equally, then $m_1 v_1^2 = m_2 v_2^2$, so $k v_1 = k v_2$,

$\therefore v_1 = v_2$ so conservation of momentum means that $m_1 = m_2$.

131. Capacitance of an isolated sphere

(a) $C = 4\pi\varepsilon_0 R = 4 \times 3.14 \times 8.85 \times 10^{-12} \times 7.5 \times 10^{-2}$
So the capacitance of the sphere is $8.34 \times 10^{-12}\,\text{F}$
$Q = CV = 8.34 \times 10^{-12} \times 1800 = 1.5 \times 10^{-8} = 15\,\text{nC}$

(b) As the spheres are identical, $C_1 = C_2 = 8.34 \times 10^{-12}\,\text{F}$. Also, we know that charge is conserved, so $Q_1 = Q_2 = \tfrac{1}{2}(1.5 \times 10^{-8})$
$= 7.5 \times 10^{-9}$ coulombs.
Using $Q = CV$, we can see that the potential at the surface of the spheres is

$V = \dfrac{Q}{C} = \dfrac{7.5 \times 10^{-9}}{8.34 \times 10^{-12}} = 900\,\text{V}$

132. Representing magnetic fields

The field lines in a diagram of a magnetic field represent the resultant magnetic force acting on a particle within that field. At any point a force can only act in one direction. If the field lines were to cross, then at the point of intersection the force would be acting in more than one direction, which is impossible. Therefore, the lines cannot cross.

133. Force on a current-carrying conductor

$F = BQv = 140 \times 10^{-3} \times 1.6 \times 10^{-19} \times 1.2 \times 10^7$
$F = 2.7 \times 10^{-13}\,\text{N}$
$F = ma$
$a = \dfrac{F}{m_e} = \dfrac{2.7 \times 10^{-13}}{9.11 \times 10^{-31}}\qquad a = 3.0 \times 10^{17}\,\text{m s}^{-1}$

134. Charged particles in a region of electric and magnetic field

(a) When an electron beam enters a uniform magnet field at right angles to the field, the electrons experience a force at right angles to their direction of travel. If all electrons have the same speed, this causes constant centripetal acceleration which, causes the beam to be deflected along a circular path.

(b) When the electron beam enters the magnetic field at an angle θ to the field, where θ is close but not equal to $90°$, the velocity of the beam can be resolved into two components, $v\sin\theta$ perpendicular to the field and $v\cos\theta$ parallel to it. Thus, the force $F = BQv$ acting on the beam can also be resolved into two components, with the component perpendicular to the field causing the beam to move in a circular direction, and the parallel component creating straight-line motion in the direction of the field. Thus the resultant force causes the beam to move along a helical path.

135. Magnetic flux and magnetic flux linkage

$\phi_1 = B.A$; $\phi_2 = B.A\cos\theta = \phi_1\cos\theta$

$\therefore \cos\theta = \dfrac{\phi_2}{\phi_1} = \dfrac{7.5 \times 10^{-4}}{1.3 \times 10^{-3}} = 0.577 \quad \therefore \theta = 55°$

136. Faraday's law of electromagnetic induction and Lenz's law

Lenz's law holds as a result of the law of conservation of energy. If the e.m.f. produced were in the same direction as the change in flux linkage (and so in the current) producing it, then the two currents would add up, creating more magnetic field, thus producing a further increase in the resultant current, and so on. This would then violate the law of conservation of energy. So the two must act in opposite directions.

137. The search coil and the simple a.c. generator

An e.m.f is induced in the coil whenever there is a change in the flux linkage. The size of the induced e.m.f. is dependent on the rate of change of flux linkage, and its direction opposes the change. From the graph, you can see that the rate of change in flux linkage is maximum when it passes through zero on the $N\phi$ axis (points A, C, and E). At A and E the flux is increasing so the induced e.m.f. will be negative; at C it will be positive. The rate of change is zero at points B and D, the maximum and minimum of the graph. Thus, we could draw the induced e.m.f on a similar curve but quarter of a cycle behind the flux linkage.

138. The ideal transformer

The law of conservation of energy is not being broken because the total energy in the system includes both the p.d. and the current induced in both coils: energy transfer $= \dfrac{V}{t}$. As the p.d. is stepped up, the current steps down in the same ratio.

140. The nuclear atom

1 (a) Force arrow should be drawn at the point on the figure where the path curves away from the gold nucleus, pointing perpendicularly away from the nucleus towards the centre of the arc.

(b) Diagram similar to the one given with the arc larger and the slope less steep.

2

Number of protons	Number of neutrons	Number of electrons	Symbol
2	2	2	^4_2He
8	8	8	$^{16}_8\text{O}$
88	138	88	$^{226}_{88}\text{Ra}$
92	146	92	$^{238}_{92}\text{U}$

141. Nuclear forces

(a) In a gold-197 nucleus of radius 7×10^{-15} m there are 197 nucleons of mass. Density (ρ) is given by:

$$\rho = \frac{m}{V}$$

Thus for the spherical nucleus:

$$\rho = \frac{A \times m_N}{\frac{4}{3}\pi R^3}$$

Where m_N is the nucleon mass, R the nuclear unit radius and A the mass number.

$$\rho = \frac{197 \times 1.67 \times 10^{-27}}{\frac{4}{3}\pi \times (7 \times 10^{-15})^3} = 2.3 \times 10^{17} \, \text{kgm}^{-3}$$

(b) Using the equation for the nuclear radius $R = r_0 A^{\frac{1}{3}}$ we see that:

$$\frac{R_{\text{Au}}}{R_{\text{Fe}}} = \left(\frac{A_{\text{Au}}}{A_{\text{Fe}}}\right)^{\frac{1}{3}}$$

$$R_{\text{Fe}} = 7 \times 10^{-15} \times \left(\frac{56}{197}\right)^{\frac{1}{3}}$$

$$= 4.6 \times 10^{-15} \, \text{m}$$

142. The Standard Model

The two types of particles that make up matter are quarks and leptons.

Quarks come in 3 'generations', each with two types of particle
1st generation: up, down
2nd generation: strange, charm
3rd generation: top, bottom

Leptons also come in 3 'generations':
1st generation: electron, electron neutrino
2nd generation: muon, muon neutrino
3rd generation: tau, tau neutrino

Antimatter is also made up from two different types of particle – antiquarks and antileptons – with the same structure of generations. Each antiparticle has the same mass but opposite charge to its matter equivalent.

143. The quark model of hadrons

Since Λ particles consist of three quarks, the Λ^0 must be a **baryon**. The Λ^0 particle is neutral. You know that the up and down quarks have charge of $+\frac{2}{3}$ and $-\frac{1}{3}$, respectively, thus in order for the particle to remain neutral, the third quark must also have a charge of $-\frac{1}{3}$. So the third quark must be one of a down, strange, or bottom quark. **The third quark is actually a strange quark, as the Λ^0 particle is a strange particle**

144. β⁻ decay and β⁺ decay in the quark model

Both types of β decay will produce more stable daughter nuclei. β^+ decay occurs when the isotope has too many protons. The result is to convert one proton into a neutron with a positron emitted. β^- decay occurs when the isotope contains too many neutrons to be stable. One of the neutrons will be converted into a proton while an electron is emitted. Most nuclei have an appropriate balance of protons and neutrons, and so beta-decay does not occur.

145. Radioactivity

When a radioactive source is used as a tracer, it is important that the detector outside the system being investigated can receive the emitted particles. As β particles are moderately ionising and therefore able to penetrate human tissue without doing too much damage to the tissue, a β source would be good for investigating blood flow. However, β particles are stopped by very thin metal and so would be undetectable from outside a system of copper pipes.

146. Balancing nuclear transformation equations

With seven α decays the isotope will lose 14 from the proton number and 28 from the nucleon number. Similarly, with four β^- decays, it will gain 4 in the proton number. So the resultant nuclide will have a proton number of 82 and a nucleon number of 207. Looking at the table of elements, this means that the final nuclide is $^{207}_{82}\text{Pb}$ – lead.

147. Radioactive decay 1

When we say that the radioactive decay of a particle is 'spontaneous', we mean that there is no external factor that causes the particle to decay. When we say it is a 'random' process, this means that there is no way of predicting the moment at which the decay will occur. What we can say is that the chance of any single particle decaying is such that between time $= t_0$ and time $= t_{\frac{1}{2}}$ 50% of the remaining particles will have decayed.

148. Radioactive decay 2

$t_{\frac{1}{2}} = 5730 \times 365.25 \times 24 \times 60 \times 60 = 1.81 \times 10^{11}$ seconds

$\lambda t_{\frac{1}{2}} = \ln 2 \Rightarrow \lambda = \dfrac{0.693}{1.81 \times 10^{11}} = 3.83 \times 10^{-12}$

The activity A_0 is given as 0.267 Bq per g. So for 2 g of carbon, $A_0 = 0.534$ Bq. Thus:

$$\frac{A}{A_0} = \frac{0.25}{0.534} = 0.468 = e^{-\lambda t}$$

So

$$t = \frac{\ln(0.468)}{-3.83 \times 10^{-12}} = 1.98 \times 10^{11} \text{ seconds}$$

So the estimated age of the wood is 6274 years.

149. Einstein's mass – energy equation

1 The electrons and positrons must be produced as (particle–antiparticle) pairs so that charge is conserved. The particle and antiparticle move outwards in opposite directions in order to conserve momentum.

2 Using Einstein's equation $\triangle E = mc^2$, given that:
$\triangle m = 0.003\,u = 0.003 \times 1.66 \times 10^{-27} = 4.98 \times 10^{-30}\,kg$ and
$c^2 = 9.00 \times 10^{16}\,m^2\,s^{-2}$
Energy released $= \triangle E = 4.48 \times 10^{-13}$ joules

150. Binding energy and binding energy per nucleon

Mass–energy is 938 MeV for each particle. Therefore total energy available for photon production is $2 \times 938 = 1876$ MeV

151. Nuclear fission

Points to include:

Fusion	Fission
Very high temperature required	
High density required to maintain fusion reaction	
More energy is produced than in the fission process	Less energy produced
Creates less radioactive waste than fission	Radioactive waste produced
Fuel supply almost unlimited as more fuel is created in the fusion process	Fuel used up in the fission process
Problems continue with containment and control of the reaction	Reaction can be controlled more easily

152. Nuclear fussion and nuclear waste

Points to include:

High level waste	Intermediate-level	Low-level
β, γ emitters. Very harmful to biological systems	Less harmful to biological systems	Still less harmful, although prolonged exposure is not advisable
Encased on site in reinforced concrete and steel	Can be reprocessed to extract useful radioisotopes	Spent fuel encapsulated in tight canisters for long-term storage
^{238}U can be converted to plutonium for use in other reactors	Transported from nuclear reactor to reprocessing plant	
Ultimately it is envisioned that all nuclear waste can be stored in 'geological repositories' – rock chambers deep in the Earth's crust – where they will no longer be harmful.		

[Students should include at least one point from each column to compare and contrast.]

154. Production of X-ray photons

The anode must be a good thermal conductor so that the heat created in the target metal when it is hit by the electrons can be conducted away from the target.
X-ray tubes are usually cooled by removing energy from the anode by circulating oil around a heat exchanger. Cooling fans to produce forced convection may also be used.

155. X-ray attenuation mechanisms

1 The standard explanation for the Compton scattering effect relies on the laws of conservation of energy and of momentum. This leads to the idea that the scattered photons must possess momentum, giving weight to the notion that they are particles rather than waves.

2 Using the attenuation equation
$$I = I_0 e^{-\mu x} \text{ with } \frac{I}{I_0} = 0.13, \text{ we have } 0.13 = e^{-0.012\mu}$$
So
$$\mu = \frac{\ln(0.13)}{-0.012} = 170\,m^{-1}$$

156. X-ray imaging and CAT scanning

CAT scanners have largely superseded techniques involving the use of barium salts because the images obtained are clearer and the detectors are much more manoeuverable than X-ray machines. Images can be obtained from many different angles without moving the patient and, because they are generally electronic images rather than (X-ray) photographs, they can be processed more fully to enhance what is being investigated.

157. Medical tracers

γ-rays are the least ionising and cannot be completely absorbed by solid materials. β-rays are much more harmful to biological tissues.

158. The gamma camera and diagnosis

• Gamma camera are used to image processes rather than structures.
• Rather than shining X-rays through the body, the detector picks up photons that are generated from an isotope emitter passing through the bloodstream.
• The image does not come through instantly, but is built up over a period of time. This enables the path of the emitter to be traced as it passes through the bloodstream, rather than as a series of 'snapshots'.

159. PET scanning diagnosis

A radioactive isotope is introduced into the area to be scanned. This isotope undergoes β^+ decay, emitting a positron. When the emitted positron encounters an electron within the tissue being scanned, an annihilation event occurs, with two gamma rays produced, travelling in opposite directions. When the PET scanner detects two gamma rays simultaneously on opposite sides of the ring, it is clear that an annihilation event has occurred; single photons can be ignored as being just stray background radiation.

160. Ultrasound

Advantages of ultrasound over X-rays
• No ionisation from sound waves – no damage to living cells
• Can be used to produce images of soft tissue while X-rays can only see bones, etc.

Disadvantages of ultrasound over X-rays
• X-rays have shorter wavelength so can create a higher resolution image
• Ultrasound images can be distorted if there is a layer of air between the subject and the detector
[students choose one from each list]

161. Acoustic impedance and the Doppler effect

(a)
- To see inside the patient, as much of the pulse needs to be transmitted through the skin as possible.
- Because the density of air is so much smaller than that of the skin, Z_1 is much smaller than Z_2, so $Z_2 - Z_1 \approx Z_2 \approx Z_2 + Z_1$.

  Thus $\dfrac{I_r}{I_o} \approx 1$. In other words, a tiny amount of the signal is transmitted (0.1% in the example).
- To make this figure larger, Z_1 and Z_2 must be much closer together, which is why a coupling gel with density similar to the skin is used. Then $\dfrac{I_r}{I_o}$ is close to 0 and a much higher percentage of the signal is transmitted through the skin.

(b) The ideal value for the acoustic impedance for the gel would be 1.64×10^6, equal to that of the skin. The, $z_1 = z_2$, so $\dfrac{I_r}{I_o} = 0$ and 100% of the signal would be transmitted through the skin

162. Ultrasound A-scan and B-scan

(a) Short pulses of small amplitude (i.e. low-energy) ultrasound waves are transmitted through the skin on the mother's abdomen. When they encounter tissues of a different density, reflection occurs at the boundary.

Signals from reflected pulses as the ultrasound transducer is moved across the skin produce dots on a screen that combine to form an image.

(b) Scanning all round produces a full image of the fetus. Ultrasound is non-invasive and because the waves are low energy, no damage is done to living cells.

Notes

Notes

Notes

Notes

Notes

Notes

Published by Pearson Education Limited, 80 Strand, London, WC2R 0RL.

www.pearsonschoolsandfecolleges.co.uk

Copies of official specifications for all OCR qualifications may be found on the OCR website: www.ocr.org.uk

Text © Pearson Education Limited 2016
Copyedited by Saskia Besier
Typeset by Tech-Set Ltd, Gateshead
Produced by Out of House Publishing
Illustrated by Tech-Set Ltd, Gateshead
Cover illustration © Miriam Sturdee

The rights of Steve Adams and Ken Clays to be identified as authors of this work have been asserted by them in accordance with the Copyright, Designs and Patents Act 1988.

First published 2016

19 18 17 16
10 9 8 7 6 5 4 3 2

British Library Cataloguing in Publication Data
A catalogue record for this book is available from the British Library

ISBN 978 1 447 98438 2

Printed in Slovakia by Neografia

Picture Credits
The publisher would like to thank the following for their kind permission to reproduce their photographs:
(Key: b-bottom; c-centre; l-left; r-right; t-top)
123RF.com: tiero 162; **Alamy Images:** MARKA 91, PHOTOTAKE Inc. 80, 156tr, sciencephotos 63tr, 63bl, 67, Tetra Images 36, Zoonar GmbH 65; **Fotolia.com:** Awe Inspiring Images 132, daboost 156tl, Roman Ivaschenko 46; **Getty Images:** AFP / Stan Honda 38, William Anderson 42; **NASA:** Kennedy Space Centre 35; **Rex Shutterstock:** Shutterstock / Quirky China News 17; **Science Photo Library Ltd:** Andrew Lambert Photography 63br, 153, Larry Mulvehill 157, Martyn F. Chillmaid 41, SPL / Detlev van Ravensway 109, Victor de Schwanberg 71; **Shutterstock.com:** Yury Dmitrienko 111

All other images © Pearson Education